Advances in Hydropower Technologies

Edited by Basel I. Abed Ismail

Published in London, United Kingdom

Advances in Hydropower Technologies
http://dx.doi.org/10.5772/intechopen.111224
Edited by Basel I. Abed Ismail

Contributors
Adam Ahmed Abdou, Ahmed Moulaye Abdou, Ana Regina Mizrahy Cuperschmid, Basel I. Abed Ismail, Ibrahim Halil Demirel, Jamal El Mansour, João Rodolfo Côrtes Pires, Moses Jeremiah Barasa Kabeyi, Oludolapo Akanni Olanrewaju, Paulo Sergio Franco Barbosa, Pedro Nogueira Resende

First published in London, United Kingdom, 2025 by IntechOpen
IntechOpen is the global imprint of INTECHOPEN LIMITED, registered in England and Wales, registration number: 11086078, 167-169 Great Portland Street, London, W1W 5PF, United Kingdom

For EU product safety concerns: IN TECH d.o.o., Prolaz Marije Krucifikse Kozulić 3, 51000 Rijeka, Croatia, info@intechopen.com or visit our website at intechopen.com.

British Library Cataloguing-in-Publication Data
A catalogue record for this book is available from the British Library

Advances in Hydropower Technologies
Edited by Basel I. Abed Ismail
p. cm.
Print ISBN 978-0-85466-026-1
Online ISBN 978-0-85466-025-4
eBook (PDF) ISBN 978-0-85466-027-8

If disposing of this product, please recycle the paper responsibly.

Meet the editor

Dr. Basel I. Abed Ismail is a successful, long-standing academician and an experienced Associate Professor and Chair of the Department of Mechanical and Mechatronics Engineering at Lakehead University. He is a pioneering international researcher in alternative energy, as well as renewable and green energy engineering technologies. Dr. Ismail serves as an academic editor for several books and has authored numerous peer-reviewed articles and book chapters. He is also a licensed Professional Engineer (P. Eng.) in Ontario, Canada, and a member of the Canadian Society for Mechanical Engineering (CSME).

Contents

Preface

Green and renewable hydropower has always been considered one of the most efficient and magnificent power generation sources. For many years, hydropower played an essential role in the development of humanity and has a long and successful track record. It is a conventional renewable energy source for generating electricity in small-, medium- and large-scale production. The power generation capacity of hydroelectric power plants can be measured by multi-gigawatt production.

This book is the result of contributions from several researchers and experts worldwide. An introductory chapter is initially presented in this book, which will highlight the principles, performance criteria, efficiency, and design correlations of major hydro-turbines used in hydroelectric power generation plants.

Due to its significant utilization and future technological prospects, various interesting topics of research related to hydroelectric power generation are covered in this book. The book covers the following interesting areas: design, operation and efficiency of hydropower plants; pumped hydropower storage systems; hydroelectric power plants and their sustainability; emergence analysis of novel hydroelectric energy markets; hydropower plants modelling, simulations and management; current and future trends of hydropower policies, and more.

I hope that the book will become a valuable source of information and a basis for extended research for researchers, academics, policymakers, and practitioners in the area of renewable hydropower technologies.

I sincerely thank all authors for their efforts and the quality of the chapters presented. I would also like to thank Ms. Iva Horvat, the book publishing process manager at IntechOpen, for her excellent efforts in managing the publication process of this book.

Dr. Basel I. Abed Ismail, P.Eng.
Associate Professors and Chair,
Department of Mechanical and Mechatronics Engineering,
Lakehead University,
Thunder Bay, Ontario, Canada

Chapter 1

Introductory Chapter: Performance and Design Aspects of Hydro-Turbines for Green Hydroelectric Power Generation

Basel I. Abed Ismail

1. Introduction

This book presents unique research topics of advances in the area of hydropower technology. Renewable hydropower is a magnificent source and one of the most important renewable sources of the world's generation of large-scale electrical power that can exceed several GW of power. It has shown a long and successful track record due to its high reliability and efficiency, competitive performance, consistency of power production, and for its reduced environmental impact as a green power source [1]. The first hydropower plant was developed in 1882 and provided an electrical power of approximately 6 kW to light bulbs. In 1883, the first known small-scale hydropower plant was developed and built in Northern Ireland to supply energy for an electric train. In the same year, a hydropower plant with a capacity of 12 kW was established in Diamantina, Brazil [2]. Hydroelectric power systems can be driven from a water stream (also known as a run-of-river) or water held behind a dam in an elevated reservoir [1]. **Figure 1** shows the growth (2017–2021) of the world's hydroelectric power-installed capacity.

The world's hydroelectric power-installed capacity reached 1360 GW in 2021, exceeding all renewable energy technologies [3]. The top hydropower producers with their respective 2021 hydropower-installed capacity (GW) are shown in **Figure 2**.

The most essential and vital component of a hydropower facility which converts the kinetic and/or pressure energy in the water flow into mechanical rotational energy (in the form of mechanical shaft work) is the hydro-turbomachinery device known as a hydro-turbine. The mechanical shaft work is then converted into electrical power, using an electrical generator, which is then regulated using electric transformers in the powerhouse of the hydropower facility, as shown in **Figure 3**. The design and innovation of hydro-turbines constitute a major part in the success of this technology. Nowadays, hydro-turbines are made more compact and usually operate at high rates of shaft rotation (RPM) with high mechanical efficiency. This introductory chapter presents and discusses fundamental performance and design aspects of three major types of hydro-turbines that are most used in hydroelectric power facilities.

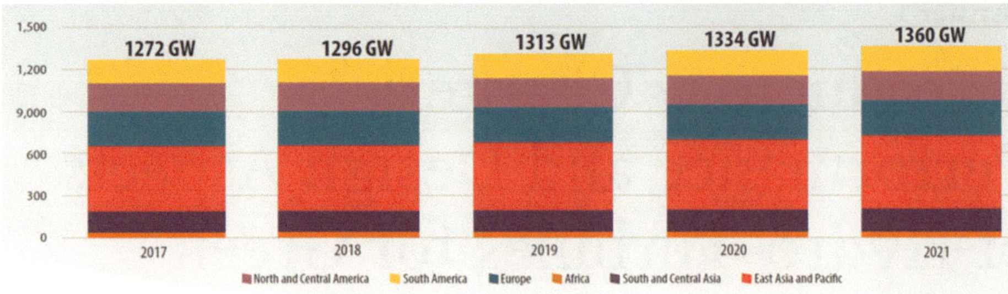

Figure 1.
The global hydroelectric power installed capacity growth, 2017–2021 (GW) [3].

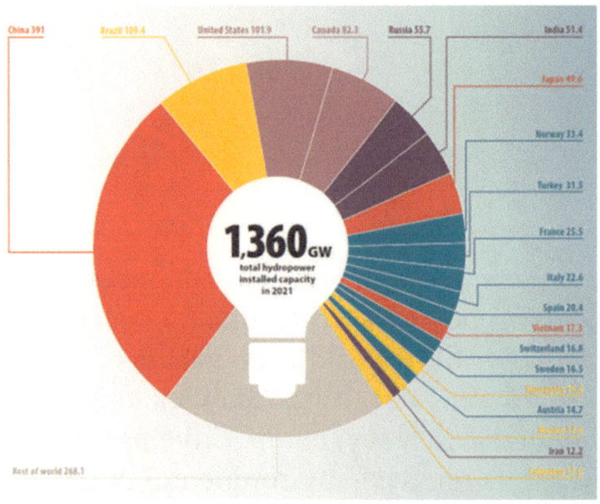

Figure 2.
The top 20 hydropower producers and the rest of the world with their respective hydropower installed capacity GW including pumped storage (2021) [3].

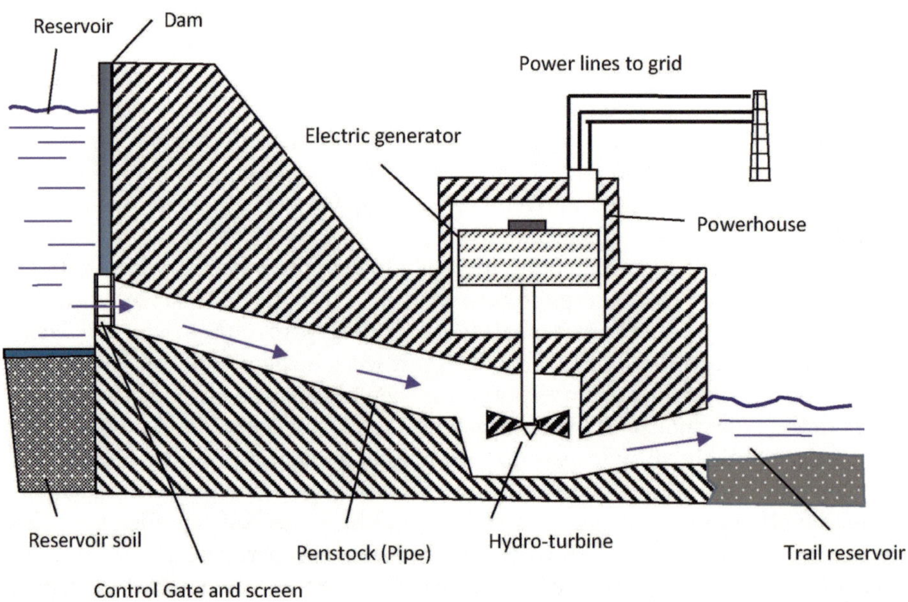

Figure 3.
A schematic diagram of a typical hydropower facility [1].

2. Types of hydro-turbines

The selection of a hydro-turbine for hydropower applications depends on the application range, primarily on the water elevation head h (m) measured between the main and trail reservoirs, the available water flow rate $\dot{\forall}_w$ (m^3/s), and other criteria, such as, maintenance, cost, sensibility to materials in suspension in the flow, and hydro-turbine mechanical efficiency. The available hydropower (W or kW) is estimated using [1]

$$P_{H,ava} = \gamma_w \dot{\forall}_w \, h \tag{1}$$

where γ_w is the specific weight for water ($\cong 9180\ N/m^3$). As a general guideline, the selection of a proper hydro-turbine to meet certain requirements of $\dot{\forall}_w$ and h can be determined using the hydro-turbine specific speed n_s given by [2].

$$n_s = n\sqrt{\frac{\dot{\forall}_w}{\sqrt{h^3}}} \tag{2}$$

In Eq. (2), n is the rotational speed of the hydro-turbine (RPM). **Table 1** shows the specific speed ranges for various types of hydro-turbines [2]. Hydro-turbines are classified as either impulse turbines or reaction turbines [4]. Reaction hydro-turbines are classified according to the angle that the flow enters the rotor blades.

In following section, three major hydro-turbines, namely, Pelton, Kaplan, and Francis, will be discussed.

2.1 The Pelton hydro-turbine

In 1878, the modern and most efficient type of impulse hydro-turbine was invented by Lester A. Pelton [4]. A schematic diagram illustrating a cross-section of a Pelton hydro-turbine is shown in **Figure 4**. It consists of the motive wheel with a diameter D which incorporates a series of vanes (buckets/shovels) protected by a housing, as shown in **Figure 4**. Simple designs of Pelton hydro-turbines allow operation under atmospheric pressure. Maintenance and repair of Pelton hydro-turbines are relatively fast due to the relatively simple construction of the rotor and its easy open access at atmospheric pressure. In recent complex designs, Pelton hydro-turbines may include multiple injectors and motive wheel of large diameter. The function of the jet

Turbine Runner	Head range (m)	n_s Range
Pelton	400–2000	0–30
Francis	50–500	20–120
Mixed-flow	20–80	120–180
Kaplan (vertical)	8–50	180–260
Bulb pit or tube (horizontal)	0–10	260–360
Michel-Banki	1.50–150	30–210

Table 1.
Recommended n_s ranges for various types of hydro-turbines [2].

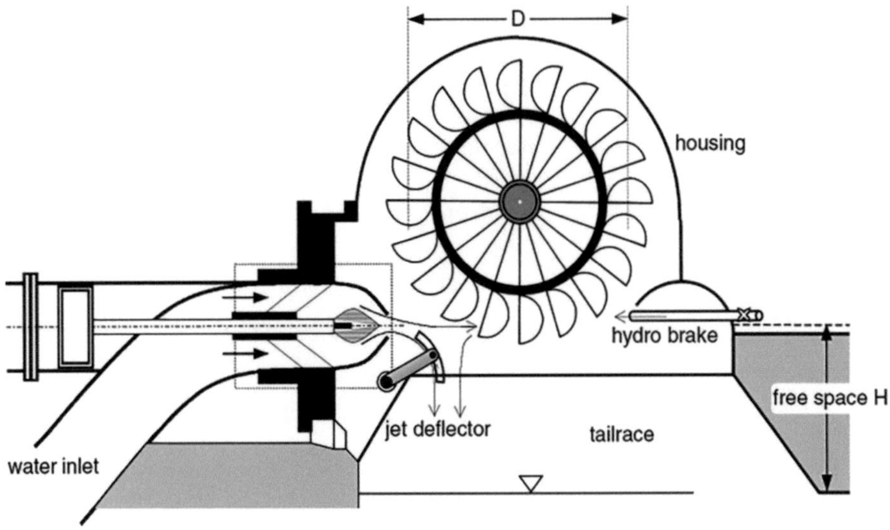

Figure 4.
A schematic diagram showing a cross section of a Pelton hydro-turbine [2].

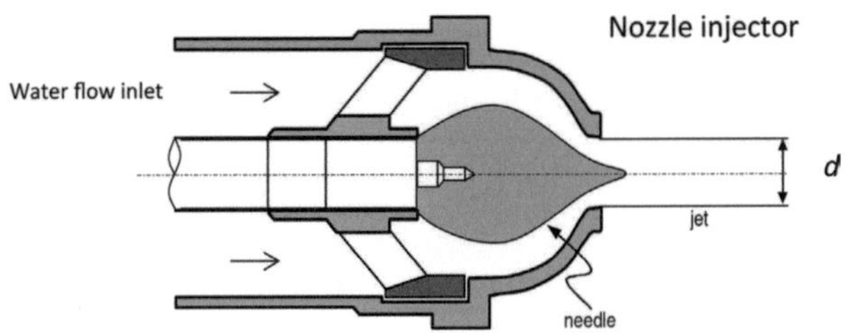

Figure 5.
A schematic diagram showing a cross section of a nozzle injector (diameter d) and needle of a Pelton hydro-turbine [2].

deflector is to control and minimize the shock intensity of the water hammer effect, a sudden reduction in the velocity of the water in the pipeline. Otherwise, a significant increase in upstream pressure may cause damages to the hydro-turbine and pipelines.

The diameter D (m) of the rotor circle at the point of jet incidence can estimated using [2]

$$D = 97.5\, \dot{\forall}_w\, (n_s)^{-0.9}\, (h)^{-0.25} \tag{3}$$

where n_s (RPM) is determined using

$$n_s = 240\, J^{0.5} \left(\frac{d}{D}\right) \tag{4}$$

where J is the number of jets used in the Peloton hydro-turbine assembly and d is the diameter of the nozzle injector, shown in **Figure 5**, given by

$$d = 0.55\, (\dot{\forall}_w)^{0.5}\, (h)^{0.25} \tag{5}$$

In order to optimize the Pelton hydro-turbine efficiency and ensure that the water exits freely after incidence of the jet on the shovels, the Mosonyi condition should be satisfied, given by

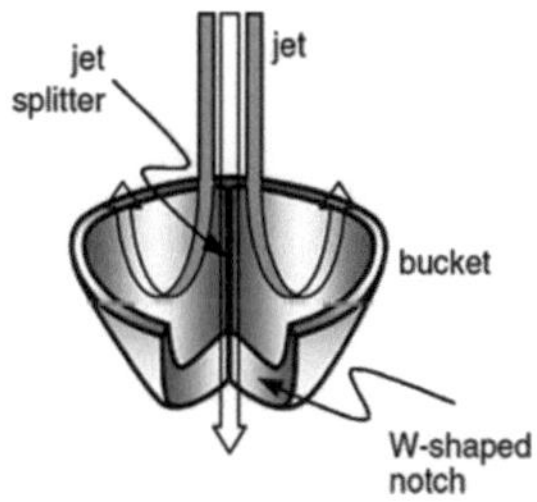

Figure 6.
A schematic diagram showing a water jet splitter [2].

$$\frac{D}{d} > 10 \tag{6}$$

The optimal diameter of the rotor circle can be obtained as

$$D_{opt} = 5.88\,(\dot{\forall}_w)^{0.5}\,(J)^{-4.5}\,(h)^{-0.25} \tag{7}$$

Most Pelton hydro-turbine designs incorporate a jet flow splitter, as shown in **Figure 6**, in order to reduce the axial thrust that would be caused by water escaping a shovel. By splitting the shovel, the water jet sustains equal force on both sides of the shovel, thus reversing the water at almost 160° to avoid striking back the incoming water. **Figure 7** shows a photograph of the detailed design of the Pelton hydro-turbine's shovels/buckets (the electric generator is shown on the right). A schematic illustration of a Pelton hydro-turbine with multiple water jet nozzles is shown in **Figure 8**.

The number of jet nozzles, J, can be estimated using the empirical formula of Tygun [5], given by

$$J = 15 + \frac{D}{2d} \tag{8}$$

2.2 The Kaplan hydro-turbine

The Kaplan hydro-turbine (also known as a propeller axial flow but with adjustable blades) was invented in 1913 by Viktor Kaplan. It is an inward reaction hydro-turbine where the flow of water enters axially (vertically) into the rotor blades. It works based on the lift force generated on the propeller blades due to its hydrofoil shape. **Figure 9** shows a photograph of a double-regulated Kaplan hydro-turbine. In this application, the water (working fluid) changes pressure (i.e., expands) as it flows through the propeller blades (at certain angles) and generates shaft work, as shown in **Figure 10**.

Unlike axial-flow propeller hydro-turbines (blades are fixed and pitch is not adjustable), Kaplan hydro-turbines are called double-regulated because the flow rate is controlled in two ways: (1) by adjusting the wicket gates (guide vanes) at inlet and (2) by adjusting the pitch on the rotor blades. Kaplan hydro-turbines are best suited for hydropower applications involving low-head and high-volume flow rate conditions with mechanical efficiencies may be as high as 94% [4].

2.3 The Francis hydro-turbine

The reaction-type Francis hydro-turbine was invented by James B. Francis in 1848 [4, 7]. **Figure 11** shows the details of a typical Francis hydro-turbine, and **Figure 12**

Figure 7.
A photograph showing the detailed design of the Pelton hydro-turbine's shovels/buckets (the electric generator is shown on the right) [4].

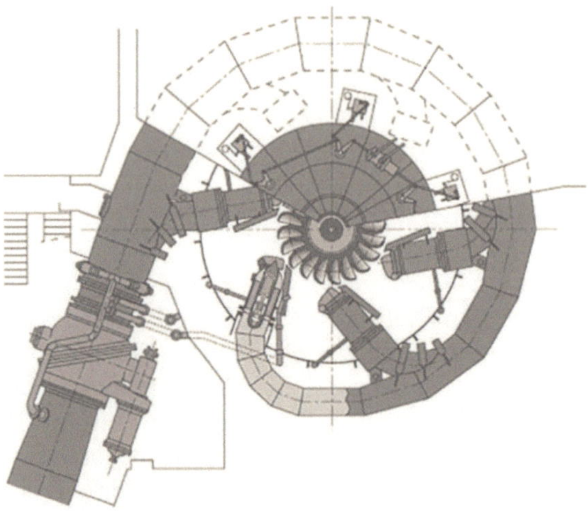

Figure 8.
An illustration showing a Pelton hydro-turbine with multiple water jet nozzles [5].

shows a photograph of the hydro-turbine. The shaft of the Francis hydro-turbine can be positioned horizontally or vertically depending on local site conditions of installing the hydro-turbine. Installing the hydro-turbine in horizontal arrangement is most suitable due the following reasons: (1) it eases direct connection of the electric generators, (2) connection in vertical arrangement may cause difficulties in repairing and maintaining the hydro-turbine, and (3) more space above the hydro-turbine is required to connect to the generator's shaft in the vertical arrangement, resulting in a

Figure 9.
A photograph showing a double-regulated Kaplan hydro-turbine [6].

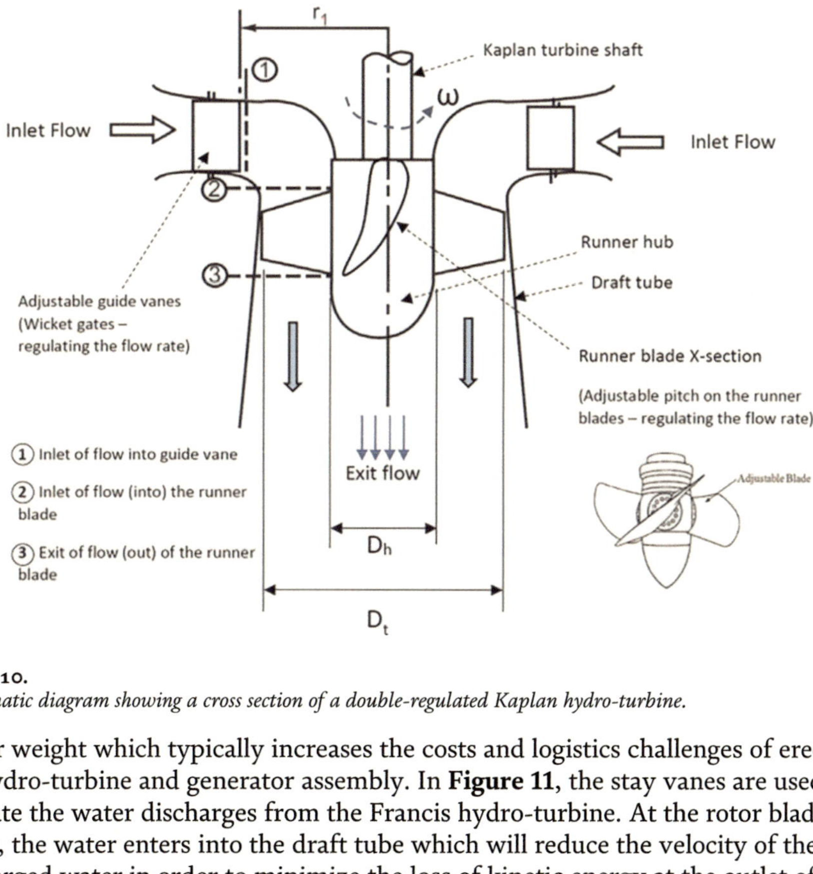

Figure 10.
A schematic diagram showing a cross section of a double-regulated Kaplan hydro-turbine.

higher weight which typically increases the costs and logistics challenges of erecting the hydro-turbine and generator assembly. In **Figure 11**, the stay vanes are used to regulate the water discharges from the Francis hydro-turbine. At the rotor blades' outlet, the water enters into the draft tube which will reduce the velocity of the discharged water in order to minimize the loss of kinetic energy at the outlet of the hydro-turbine.

The diameter of the rotor (runner) of a Francis hydro-turbine can be determined using Bereshoni's empirical correlation, for head range 10–50 m, which is given by [2]

$$D = \frac{[35.10 + 0.16\, n_s]\, (h)^{0.5}}{n} \tag{9}$$

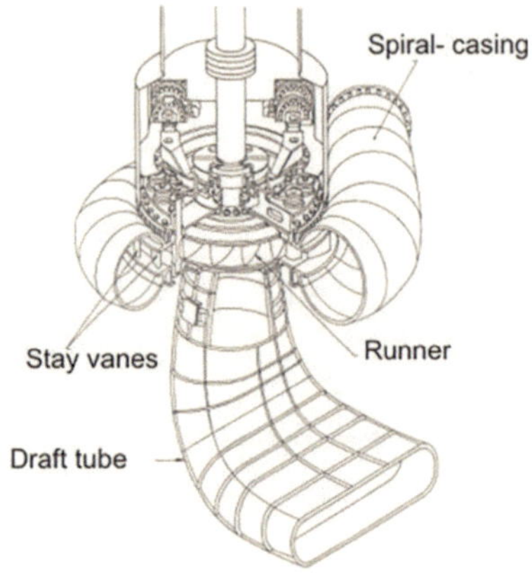

Figure 11.
A schematic diagram showing the details of a Francis hydro-turbine [8].

Figure 12.
A photograph showing the detailed design of a 230 MW Francis hydro-turbine's 17-blade rotor [4].

where n is the rotor's rotational speed in the range 900–1200 RPM and n_s is the Francis hydro-turbine specific speed given by [2]

$$n_s = \frac{n\sqrt{(1.36\,P_T)}}{h^{1.25}} \tag{10}$$

In Eq. (10), P_T (kW) is the power output of the Francis hydro-turbine. For heads higher than 50 m, the range of n is 1200–1500 RPM.

2.4 Maximum efficiency of hydro-turbines

Hydro-turbines vary in terms of their performance efficiency. **Figure 13** shows the maximum efficiency of the three hydro-turbines. N_{st} is the dimensionless specific speed of the hydro-turbine, given by [4]

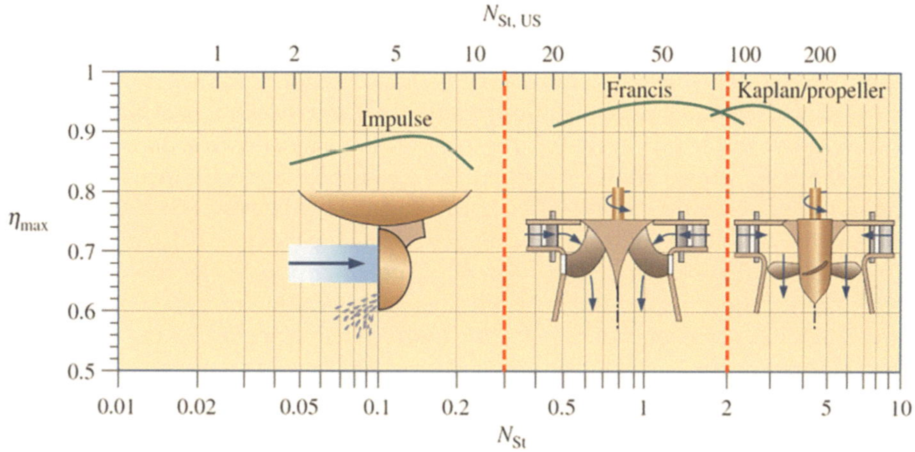

Figure 13.
Maximum efficiency as a function of N_s for the three main types of turbomachines [4].

$$N_{st} = \frac{\omega\sqrt{Bhp}}{\rho^{0.5}(gh)^{1.25}} \tag{11}$$

In Eq. (11), ω is the rotational speed (rad/s), ρ is water density (kg/m^3), g is gravitational acceleration constant (9.81 m/s^2), and *Bhp* is the brake horsepower/shaft power (W). As shown in **Figure 13**, impulse Pelton hydro-turbines perform optimally at N_{st} value near 0.15, while reaction Francis and Kaplan hydro-turbines perform best at N_{st} near 1 and 2.5, respectively.

3. Specifications and challenges

In a hydropower project involving the selection and operation of a hydro-turbine coupled with an electric generator, it is necessary to specify the following [2]:

1. The available head (minimum, maximum, and average) and water flow rate at the potential hydropower site

2. Water quality in terms of material in suspension and abrasiveness

3. Present and future electrical energy demands taking into consideration the effective capacity of the hydropower site

4. Characteristics (e.g., type and output, efficiency, etc.) of the electric generator

5. Location of the powerhouse.

It was reported [7] that the most common reasons for damages to hydro-turbines, specifically Kaplan, are cavitation, silt erosion, and dynamic fatigue. The erosion (also known as hydro-abrasive wear) of hydro-turbines due to suspension of material in water flow (e.g., solid and hard particles) and cavitation (clearance tip cavitation and tip vortex cavitation) has been challenging in the design, construction, operation, and maintenance of hydropower plants [7]. Cavitation inside the hydro-turbine may occur due to the fact that the pressure frequently drops below the saturation vapor pressure of water at the prevailing temperature. It could potentially result in several issues, namely, vibration, shock waves, noise emission, and micro jets. It was also reported that when a hydro-turbine in the hydropower facility shuts down suddenly, and it could potentially damage the moving parts and any operating equipment and instruments connected to them [7, 9]. Pelton hydro-turbines are subject to erosion from the fluid along its central jet splitter resulting in performance losses [10]. This highlights the significance of maintaining the water jet splitter sharp to be effective in maximizing efficiency. Deterioration of the functioning and operation of hydropower plants causes disastrous problems not only for the hydro-turbine performance, but may lead to total collapse of the hydropower plant. The various types of damages, types of failure mechanisms, and suggested methods to prevent them in hydro-turbines are extensively discussed in Refs. [7, 9]. More details related to hydropower aspects are given in Ref. [1].

Author details

Basel I. Abed Ismail
Department of Mechanical and Mechatronics Engineering, Lakehead University, Canada

*Address all correspondence to: baseliai@gmail.com

References

[1] Ismail BI. Aspects of renewable hydroelectric power generation, introductory chapter. In: Ismail BI, editor. Renewable Hydropower Technologies. Rijeka, Croatia: InTech Open Access Publishing Company; 2017. eBook ISBN: 978-953-51-4723-7

[2] Farret FA, Simoes MG. Integration of Alternative Sources of Energy. New Jersey, U.S.A: John Wiley & Sons, Inc.; 2006. ISBN: 978-0-471-71232-9

[3] International Hydropower Association (IHA). The 2022 Hydropower Status Report: Sector Trends and Insights. England: International Hydropower Association (IHA); 2022

[4] Kanoglu M et al. Fundamentals and Applications of Renewable Energy. U.S.A.: McGraw-Hill; 2020. ISBN: 978-1-260-45530-4

[5] Korpela SA. Principles of Turbomachinery. 2nd ed. New Jersey, U.S.A: John Wiley & Sons, Inc.; 2020. eBook ISBN: 978-1-119-51811-2

[6] Kaplan Efficiency Improvement System - REIVAX. Available from: https://www.reivax.com/us/automatic-cam-optimizing-system-for-kaplan-and-bulb-turbines/ [Accessed: May 21, 2024]

[7] Ridha WKM et al. Common failures in hydraulic Kaplan turbine blades and practical solutions. Materials. 2023; **16**(3303):1-37

[8] Quora. Available from: https://www.quora.com/What-is-Francis-turbine [Accessed: May 21, 2024]

[9] Dorji U, Ghomashchi R. Hydro turbine failure mechanisms: An overview. Engineering Failure Analysis. 2014;**44**:136-147

[10] Aggidis G et al. Methodology for the optimization of a novel hydro turbine with a case study. Energies. 2023; **16**(7591):1-17

Chapter 2

Design and Operation of Hydropower Plants with BIM Digital Technologies

João Rodolfo Côrtes Pires, Paulo Sergio Franco Barbosa, Pedro Nogueira Resende and Ana Regina Mizrahy Cuperschmid

Abstract

The intricate nature of hydropower plant design and operation, coupled with multiple domains of expertise, regulations, and numerous stakeholders, presents prospects for enhancing quality and cutting overall costs through cooperation and integration. The practical application of Virtual Design and Construction (VDC) and Building Information Modeling (BIM) in the design and O&M planning stages can significantly contribute to these goals. This chapter's material is partially based on firsthand observations made while working on an R&D project funded by a consortium of hydropower-producing enterprises. To enhance the quality of the project life cycle as a whole, we plan to draw attention to particular BIM modeling concerns that ought to be considered while designing hydropower components (such as water intakes and powerhouses). The case study is based on the BIM modeling of three built plants to construct a digital twin. These HPPs are in South Brazil: (1) Barra Grande hydropower plant, with an installed capacity of 690 MW; (2) Foz do Chapecó hydropower plant, with an installed capacity of 855 MW; (3) 14 de Julho hydropower plant, with an installed capacity of 100 MW.

Keywords: Building Information Modeling, BIM hydropower cases, digital technologies, Virtual Design and Construction, 3D modeling

1. Introduction

Hydropower plants are valuable infrastructure assets that benefit society. They are a clean source of mature technology with a portfolio of relevant operational attributes significantly contributing to improved grid operation. Nowadays, these characteristics are reinforced in markets where renewable and distributed generation has grown.

As an asset that must be tailored to local conditions, the design tasks of hydropower plants are highly complex, requiring precision, effective coordination, and integration of several disciplines, including geological and geotechnical, structural, electromechanical, structural, hydraulics and hydrology, and environmental sciences, among other areas. In addition to complexities of the design stage, the overall process along life cycle is also challenging, as it encompasses planning, construction, operation, and maintenance.

These challenges have now been more effectively addressed with the technological advancement of Building Information Modeling (BIM) and Virtual Design and Construction (VDC). BIM is a dominant set of technologies that embraces the multiple stages of design, construction, and management [1]. Although it is a little late compared to other application areas, BIM has progressively started to be assimilated from the hydropower industry.

More than a simple technological tool, BIM and VDC have impacted the whole process of engineering concepts and practices for design thinking as it allows exploiting various layout and design alternatives in a much faster way. The accuracy of the design reduces construction execution errors and contributes to have more precise cost estimates. Furthermore, these advances contributed to review conventional procedures of procurement and contracting along the supply chain of service providers, who can have better cost estimates and avoid unexpected changes during the construction phase.

2. The basic of BIM concepts and virtual design and construction

VDC and BIM are increasingly recognized as transformative approaches in the construction industry, enhancing project communication, coordination, and overall management [2]. These methodologies are being integrated into various stages of construction projects to achieve digital construction implementation, improving outcomes in terms of efficiency, reduced waste, and time management.

VDC refers to technologies that aid communication, visualization, and organizational design in construction projects [3]. It is a multi-disciplinary approach that utilizes integrated performance models to support explicit business objectives in design-construction projects [4, 5]. Facilitating communication and visualizing buildability and construction sequences impacts project success in terms of delivery time and cost [4]. VDC is part of integrated project delivery, necessitating a collaborative approach and buy-in from all involved parties, including architects, engineers, contractors, and owners [6].

VDC is applied in all stages of construction projects, including bidding, project feasibility study, and construction management, enabling visualization of construction status and expenditures over time and improving efficiency, speed, and accuracy [4, 7]. It involves designing for prefabrication, conducting constructability reviews, and automating field layout and condition assessments [8], enabling the integration of various dimensions of information, such as time scheduling and cost estimation, with a 3D model to simulate the construction process [4]. So, it serves as the primary link between the architectural phase and the planning and scheduling phase of construction projects [2].

The integration of VDC with BIM impacts critical success factors in construction projects. It is important to note that this integration requires a specific team structure, software, and production ecosystem for effective implementation.

BIM stands as an essential element to the broader Industry 4.0 initiative within the construction sector. The most widespread concept of BIM is a set of policies, processes, and technologies that, when combined, generate a methodology for managing the process of designing a building or facility and simulating its performance, managing its information and data, and using digital platforms, throughout its lifecycle [9].

It has been labeled as a solution, approach, or methodology, depending on the context of the investigation; nonetheless, it is notably linked with numerous technologies [10].

The benefits of VDC and BIM have been proven in various fields of Architecture, Engineering, Construction, and Operation (AECO). During construction, these technologies can reduce time, cost, design conflicts, requests for information, change orders, project duration, and improve operational efficiency [11–14]. In project management, BIM improves cost estimation accuracy, simplifies the detection of cost-related risks, and aids in decision-making [15]. Model-based tools have been also developed for building performance analysis, such as safety risk assessment during the design phase of construction projects [16] and support in the sustainable retrofit process of buildings [17].

There are also applications for asset tracking, maintenance planning, space management, and performance monitoring. This can be achieved through integration with concepts such as the Internet of Things (IoT) [18], using sensors, or with virtual reality (VR) [19]. There is also an effort for internal integration between BIM software, streamlining engineering processes, reducing design errors, and ensuring consistency among projects [20].

The ISO 19650 series standard is intended to offer guidelines and requirements for effectively managing information throughout the entire life cycle of a constructed asset, including BIM. One important issue covered in this standard set is the use of CDE (Common Data Environment). The CDE has been improved significantly in the last few years and is probably the key point to the adoption of BIM and VDC in the HPP and infrastructure projects in general. The CDE solves, or at least decreases, the problem of exchange of work of different teams, version control, clash detection, and quality control of the models.

3. BIM and VDC as a transformative framework in hydropower design and O&M tasks

Hydropower design and operation is a fascinating area of engineering with more than 100 years of experience, continuing learning from best international practices, and scientific and technological evolution. Despite the significant advances and milestones already achieved, it is still recognized as a challenging task derived from many factors as follows:

a. The design of a hydropower plant is a complex decision-making process encompassing many disciplines and addressing a huge number of components.

b. Hydropower development is an endeavor not fully amenable to standardization, as highlighted by [21] since the project is customized for each site, requiring specific facilities which must functional and economically sound, including the dam, turbine, intake, spillway, and other components.

c. There is strong interplay among the design components, requiring many revisions of the layout design and progressive refinements along the design cycle, focusing on a compromise solution between potential benefits (production of the plant in MWh, capacity MW, and other reservoir uses), costs (CAPEX and O&M), safety issues, and environmental aspects.

d. The choice of a design parameter and the overall hydropower project design can impact the functional performance, safety, and costs along the entire asset life calendar, which has a long time duration, over 100 years in many cases.

The factors above altogether have been the basis for the growing applications of BIM resources, both in the design as well as operation of hydropower plants across the world, considering the following BIM/virtual design resources:

a. BIM models can deal with the design complexities of every discipline, encompassing: (a.1) civil engineering involved in the hydropower projects with tools to place the machinery in exact position, the correct 3D spatial integration of water intake, tunnels, and pipelines, and powerhouse, among other tasks that require precision; (a.2) mechanical engineering involved in the design, manufacture, and selection of the turbine, bearings, gears, valves, governors, etc.; (a.3) electrical engineering related to the design, manufacture, and selection of the generators, control systems, transformers, etc.

b. Despite hydropower plants' customized requirements and highly dependent on-site conditions, BIM resources have automatized the creation and representation of most design components modularly and allow their integration into a common platform.

c. The interplay among the design choices, structure by structure, can be easily simulated by reconfiguration, allowing an accurate estimation of volumes (concrete for the dam, intake, spillway; earth excavations, tunnels, etc.), pipeline lengths, water storage, and further items which impact the costs. These tools allow us to explore many design options faster than older methodologies. Therefore, the benefits of higher scope variety and functional outputs of a hydropower plant and possible lower costs are clear values toward adopting BIM/virtual design of hydropower plants.

As highlighted in [22], the implementation of BIM-based hydropower projects has increased in recent years to manage project complexity, variability, and uncertainty in hydropower engineering. This helps manage risk, information, and resources.

BIM can be used for mapping 3D terrain modeling, subsurface modeling, 3D collaborative design, costing and scheduling, visualization, and clash decision-making during the planning and design phase. It can also be used for the construction stage information repository, progress tracking, safety control, coordination among specialists, and simulation of building activities [23].

It is important to emphasize the expanding BIM and virtual reality applications as essential tools for asset management in addition to employing BIM throughout the planning and design phases. According to Ye et al. [23], the BIM resources can store information such as specifications, manuals, equipment conditions, and historical maintenance. This allows for the visualization of the operation and maintenance of electromechanical and hydromechanical equipment during the operation stage.

To maximize equipment performance and service life, the plant must be operated efficiently and undergo regular maintenance and upgrades. Given the projection that half of the current power plants will need to be upgraded by 2030 [24] and the difficulty of keeping hydropower plants' energy costs competitive with other energy sources like solar and wind [25], hydropower plant operation and maintenance must change, based on digital technologies.

4. Selected cases of BIM technology applied to hydropower

The applications presented herein aim to illustrate the successful use of BIM modeling and digital technologies. The scope of applications is large, encompassing huge, medium, and small plants, a variety of dam types (rockfill, earth, etc.), pumped storage hydro, run-off-river and regulating capacities, and multipurpose hydropower plants. We selected just a few cases to illustrate the benefits of digital tools.

The Arizona Glen Canyon Dam, which supplies water and electricity to the western United States, is the first case to be discussed. The U.S. Bureau of Reclamation, the asset's owner, sought to evaluate the dam's condition as it needed more frequent maintenance because it is already 40 years old. For this, 3D modeling and reality capture were employed. A 3D model gave the U.S. Bureau of Reclamation a new set of tools to help manage and maintain Glen Canyon Dam, as detailed in [26]. It would simplify locating, identifying, and fixing existing damage, forecasting, and minimizing future damage risks.

The U.S. Bureau's initial step was data collection. Realistic data capture from a large site entailed gathering information about the hydropower plant and dam, including measurements and images and assembling it into a computer model. They collected data with sonar, cameras, and laser-based scanners to fully document the location.

The second step was converting the gathered data into digital models. The Glen Canyon scans recorded millions of data points in rich, photorealistic 3D point clouds, producing high quality images. The team employed ReCap Pro [26] to combine this raw data into an immense set, thorough point cloud.

In the third phase, the group constructed a three-dimensional technical model and filled it in with each dam component, including the hydropower generators and structural steel. This model integrated the old 2D drawings with the more recent point cloud data to help users understand not just the individual components of the dam but also how they work together as a system—a crucial resource for BIM resources.

In the final step, to complete the project, the team integrated the technical model into InfraWorks and added real-time data streams regarding the dam's performance, such as the amount of water released and energy generated. The ultimate result is a dynamic virtual model of the dam that operators can use to pinpoint potential growth and dangers that need to be managed.

As a second case, we took the Longkaikou Hydropower Station in China's Yunnan Province, located inside the Jinsha River Basin, China. The PowerChina Huadong Engineering Corporation (PCHEC) owns a 1800 MW installed concrete gravity dam that is part of the station. Bentley [27] claims that this hydroelectric project was the first in China to fully employ 3D digital design technology across all significant design domains, including geology, hub, and plant.

A complex topography and dam foundation with deep groves, not typical in hydropower engineering, called for comparing of design and construction scenarios with 3D modeling.

Bentley Map and Decartes were used to simulate the physical geology conditions, and optioneering methods were employed to identify the best place for the dam and its layout. Using GEOPAK and MicroStation, designs for the dam foundation's excavation surface and deep grooves were carried out.

Thanks to the 3D digital design technology, the project was finished 10 months ahead of schedule. This technology also enhanced communication between construction and design and shortened the design cycle. The platform was customized to enhance product quality and minimize 3D design inaccuracies and collisions.

Based on the management environment established for this project, Lidar [28] claims that PHEC developed its own standard for future projects. The business used ProjectWise to manage configuration based on project similarities to automatically align procedures for practical collaborative BIM standards. This improved and expedited real-time information exchange.

A third case is a small installed capacity, the Smisto project, in Norway. This project was produced on the BIM platform from the conceptual planning stage to the final stage [22]. The team claims to be one of the success stories in constructing hydropower projects without issuing traditional printed 2D drawings as they used the BIM platform to disseminate the information digitally.

The Smisto project, according to Smith and Hansen [29], consists of two high-head power plants in Nordland, northern Norway: Smibelg, which has a capacity of 33 MW, and Storåvatn, which has a capacity of 25 + 8 MW. The hydropower generation was completed in three stages in 2019 after building work began in the summer of 2015. The plant is located in a remote coastal region with notable differences in temperature, precipitation, and wind patterns between the fjord and the mountains. Most of the intake sites are only reachable by helicopter, making the site accessible only by water, which places a heavy strain on execution logistics and planning.

The consulting firm has replaced traditional design techniques with cutting-edge 3D/BIM technologies, including Autodesk Revit, Civil 3D, Solibri, and Navisworks, along with enhancements to the planning and design process, to eliminate traditional printed 2D drawings [30].

The Smisto project's success story has convinced the primary contractor, Haehre Entreprenør AS, about the value of drawingless execution. The model's seamless data flow has streamlined communication between the design team, contractor, and project owner, enabling them to identify and rectify costly errors before construction commences. Moreover, the simplicity of scheduling and sequencing the design and construction processes using models has made it effortless to monitor progress by defining statuses.

All components, including building joints, for the Smisto project's concrete work, were modeled per the intended casting sequence. This provides a precise foundation for design work, including effects on structural calculations and accurate length and placement specifications for reinforcing bars using various colors. It permits early construction planning and procurement as an additional benefit for the contractor through direct quantification.

Another illustrative case is the Wuqiangxi Hydroelectric Power Station Expansion Project in China, controlled by POWERCHINA Zhongnan Engineering Corporation Limited (PZEC). Its current 1200 MW capacity has increased by 500 MW with the expansion. Bentley Expert [27] states that this project included difficult construction conditions, substantial engineering quantities, a convoluted pipeline, and difficulty managing a multidisciplinary workforce under pressure. POWERCHINA ZhongNan offers engineering survey, design, and digital construction services. They understood that integrated 3D BIM approaches were more critical than previous manual design strategies to accomplish lifecycle digital plant management.

The main challenges of the Wuqiangxi plant: the high risk of on-site control, the high requirement for construction quality, limited working progress, the accuracy requirement of work volume, the duty of environment protection, and the complexity of the pipeline system. Aiming to address those challenges, they created one platform, one data center, one set of standards, one workflow, and one time delivery system, thus integrating all parties (client, supervisor, designer, and construction) into management through all project stages.

The design of the structures for the 500 MW expansion was a task undertaken by Bentley's digital resources. The selection of ContextCapture to generate a reality mesh was a key step in this process. Bentley's open 3D modeling applications, including OpenPlant, were instrumental in establishing a BIM design model and digital twin. The use of ProjectWise facilitated seamless data and model sharing among 13 disciplines. These applications not only helped decrease design time by 2 months but also resolved over 70 design issues. The digital application, with its strong integration, provided visual management and control throughout the construction process.

Table 1 presents other illustrative cases encompassing hydropower plants in several countries and continents, ranging from small/medium-sized to large-scale projects.

Case/basic technical information	Purpose of the digital modeling	Digital tools
Xiluodu Hydropower Station, China. 13,860 MW	Digital system for asset management of the hydropower plant for the entire life cycle.	Bentley platform / Digital iTwin and ProjectWise, collaborative 3D modeling.
Datengxia Water Conservancy Hydropower Project; 1600 MW	Initial studies and complete design of the hydropower plant.	Bentley Platform, 3-D modeling/ cooperative design management. Use of Micro Station, Open Roads Designer and Open Buildings.
Guanyinyan Hydropower Station, Jinsha river, Yunnan, China; 3000 MW	Design and support to construction phase. Owners using the 3D BIM Construction System with Autodesk BIM software for procedures in construction, providing understanding of the engineering construction progress and improve the level of construction management.	Autodesk Tools: BIM 360™, BIM 360™ Glue Revit®, Revit® MEP, Navisworks® 3D Browsing Solution for Windows System – Autodesk Navisworks, iPad clients are provided with a simulation solution for 3D construction. Obs. BIM 360 tools are part of ACC (Autodesk Construction Claud).
Skavica hydropower project, Albania. 210 MW	Preliminary design of the plant. Also, front-end-engineering and design study that will contribute to the project's cost estimate.	A comprehensive BIM structure will be used to manage the project throughout the delivery process. Standard LiDAR, photogrammetry, and other digital data collection techniques will be used to support the development of the scheme proposals.
Nant de Drance Pumped Storage Hydro, 900 MW Switzerland	Powerhouse Cavern Design with BIM resources used by AFRY, the consultant company	BIM tools used to support coordination and information sharing among owners, planners and contractors. The project won the Swiss Arc BIM Award 2017.
Schils power plant, Swisitzerland, a 100 year old plant, 14 MW	Support to renewal and upgrade the hydropower plant	BIM with 60 digital sub-models, all georeferenced in cloud, weekly updates during the design phase. BIM2Field during the construction phase.

Case/basic technical information	**Purpose of the digital modeling**	**Digital tools**
Upper Arun hydropower (UAHEP), Nepal, 1061 MW	BIM for the construction, to streamline the design process and enhance coordination among different disciplines and stakeholders.	BIM tools and digital twin of the hydroelectric facility, to facilitate management and maintenance tasks.
Abdelmoumen pumped-storage hydro power plant, 350 MW, Morocco	BIM as a support tool for design of the plant.	Autodesk Revit integrating mechanical files made by Siemens NX through .sat files and Inventor. Integration of all models into Navisworks through Revit to be able to do classic 3D coordination.
Keeyask Hydroelectric Generating Station, Nelson river, Canada, 693 MW	Digital tools/BIM as support for design and plant tests.	Bentley's MicroStation software; AECOsim Building Designer, Navigator, ProjectWise.
Foz Tua hydropower plant, Portugal, 264 MW	Definition of an accurate geometric model of the site/dam (double-curved arch dam; hydraulic system with two underground tunnels.)	Bentley MicroStation and OpenRoads. Coordination of the dam drawings in MicroStation.

Table 1.
Cases of BIM application.

5. Description of BIM applications in selected hydropower plants: Foz do Chapecó, Barra Grande, and 14 de Julho

BIM modeling of the Foz do Chapecó, Barra Grande, and 14 de Julho hydroelectric plants, all of which are already constructed, was developed with the objective of creating digital twins of these structures. The initial goal of the digital twin was the structural monitoring of the plants, but the models were also designed to be used for maintenance and asset management.

The Foz do Chapecó plant has an installed capacity of 855 MW, which is divided among four Francis-type turbines. Additionally, it features a rockfill dam with a length of 598 meters and a height of 48 meters, as well as a spillway with 15 gates. Construction began in 2006 and was completed in 2011, marking it as the first in Brazil to use an asphalt core in the dam.

In turn, the Barra Grande plant, initiated in 2001 and completed in 2006, has three Francis-type turbines, each with a capacity of 230 MW. It is notable for its dam, which is 185 meters high and 665 meters long, constructed with rockfill and a concrete face. Additionally, its spillway features six large gates, each 15 meters wide and 20.98 meters high.

The 14 de Julho plant, constructed between 2004 and 2009, has an installed capacity of 100 MW and features Kaplan-type turbines. The dam was built using roller-compacted concrete and designed based on a gravity system. Another notable aspect is that its powerhouse and many galleries were excavated into rock. In **Figure 1**, the geographic positions of the power plants and their structures are compared to their respective models.

Hydroelectric power plants are complex structures of various disciplines, requiring constant monitoring and maintenance throughout their operational lifespan. So, to efficiently build a digital twin model that meets the needs of scalability, interoperability, and persistence, it was decided to adopt the best market practices regarding

Figure 1.
Foz do Chapecó, Barra Grande, and 14 de Julho hydroelectric plants.

standards and norms. This included following the data management practices outlined in the ISO 19650 series, standardizing elements according to the categorization of IFC 4.1, and structuring data according to the COBie (Construction Operations Building Information Exchange) standard.

From a software standpoint, Autodesk tools were chosen. The BIM Collaborate Pro module from Autodesk Construction Cloud was used as the CDE (Common Data Environment) for model files and data management, AutoCAD Civil 3D for terrain modeling, Revit for civil, electrical and mechanical modeling, and Recap Pro for point cloud processing. AutoCAD was also used to open As-Is projects related to the structures. Additionally, a digital twin web system using Autodesk platform services (formerly Forge), named DIGIDAM, was used.

As the three participating power plants in this project were well-documented and had high-quality digitalized drawings and technical specifications, the decision was made to start the modeling process from the 2D drawings. After completing the models, laser scanning was adopted to ensure that the model accurately corresponds to the real situation for the most critical areas with more systems, such as power houses and control rooms. **Figure 2** shows examples of point clouds for a high-voltage control room and a transformer in a substation.

A spatial and discipline-based federation approach was adopted to facilitate the modeling of the various disciplines and their integration, leading to clash detection. Generally, in a hydroelectric power plant (HPP), the following substructures were adopted for spatial federation: dams, spillways, powerhouses, tunnels, and intake.

Each substructure passed through a crucial stage in terms of work and data organization, which involved subdivision into floors and spaces (functional areas). The separation by disciplines was carried out in the following domains: site survey, civil engineering, mechanical, electrical, and monitoring system. **Figure 3** shows the representation of disciplines and types of systems addressed in the modeling.

Among the observed benefits of this separation are improved navigation within the project, easier collaboration within the modeling team, simplified version control of files, and reduced hardware requirements. Additionally, this approach may facilitate necessary future specific modifications to the model.

Figure 2.
Control room and transformer point clouds.

Site Survey	Civil	Structural Monitoring Instrumentation	Electrical	Mechanical	
				Equipaments	Systems
Bathymetry	Access Tunnels	3D Crack Meters	Emergency Generator	Turbines	Air Conditioning System
Topography	Cofferdams and Dikes	Drains	High Voltage System	Gates	Fire Protection System
	Conventional Buildings	Extensometers	Lighting	Generators	Lubricating and Insulating
	Dam Blocks	Flowmeters	Lightning Protection System	Mechanical Handling	Oil System
	Drainage Tunnels	Inclinometers	Low Voltage System		Sanitary Sewage System
	Open Channel	Penduluns	Telecommunications and		Service Compressed Air System
	Penstock	Piezometers	CCTV System		Treated Water System
	Plinth	Prisms			Ventilation System
	Spillway Crest	Settlement Gauges			Water-Oil Separation and
	Spillways	Survey Markers			Drainage System
	Stilling Basin	Tensiometers			
	Tail Race	Thermometers			
	Trunnion Beams	Total Earth Pressure Cells			

Figure 3.
Disciplines and types of systems.

Finally, another important consideration at this stage was the precise geographic location survey of the structures and their georeferencing in the modeling software. This was done to facilitate the integration of the model with GIS databases and ensure the federation of models on the collaboration platform without the need for manual interventions.

5.1 Site survey

In the cited cases, the topography and bathymetry used were derived from the plant's original construction projects, where natural contour lines were integrated with the earthworks documented in the consulted As-Is projects. The fundamental purpose of the topography was to promote precise visual identification and compatibility of all involved structures (**Figure 4**). In addition to the geometric representation, renders of the terrain were obtained from recent aerial photos to enhance the visualization of the scenario.

Although photogrammetry, laser scanning, and satellite imagery technologies were not employed for this discipline, their use is crucial for validating and creating a more

Figure 4.
14 de Julho hydroeletric powerplant federated model.

Figure 5.
14 de Julho hydropower plant topography.

robust and detailed terrain model. Furthermore, considering that the topographic model can potentially serve as a basis for developing hydrological and hydraulic simulation models, incorporating these techniques becomes even more promising.

Still, regarding the terrain model, it is important to include elements that facilitate georeferencing, such as a North-East grid with intervals 100 meters and references axes for structures. In **Figure 5**, we can see the digital terrain modeling, the contour lines extracted from the projects, the East and North grid, as well as the main axis of the 14 de Julho Plant dam, all surveyed by UTM zone 22S—SIRGAS 2000 coordinate system.

5.2 Civil engineering design

While modeling elements in the civil discipline, the aim was to encompass both geometric and material aspects of each element. Conventional buildings such as office buildings, metal roofs, walls, and partitions followed the BIM standards. Additionally, detailed structural connections or reinforcement were not included. Their finished total thicknesses were considered for floors and walls, disregarding the detailed thickness of structural and finishing layers.

The main premise was to create a base for the subsequent positioning of electro-mechanical equipment and instruments in a way that is compatible with the structure. Although there was no detailed information regarding other parameters, such as the mentioned reinforcement, the project remains faithful to the actual structure.

Some specific details considered relevant for the digital model are included for specific hydraulic structures. For example, dams are divided into blocks and layers, and this information is inserted during the modeling process. Therefore, in addition to following the complex geometry that these structures present, they are also divided into layers and blocks to better understand their internal structure. Illustrating the usage of this information allows elements such as expansion joints to be easily located in the design.

Another important factor is the subdivision of structures into "floors" and "spaces," as depicted in **Figures 6** and **7** for a powerhouse structure. This division facilitates the categorization of elements to be modeled, allowing for greater detail in their location information. An example application specifies equipment by its sector when creating a maintenance ticket; these data will already be linked through the model.

In the case of the concrete blocks of spillways, the blocks were modeled as whole pieces considering their expansion joints. The precast pieces of tunnels and galleries were modeled as separate elements from the overall structure, with their materials properly differentiated. In the spillway block of the Foz do Chapecó hydropower plant (**Figure 8**), it is possible to see in the galleries that the precast concrete voussoirs and the subsequently concreted floor are distinct elements.

In modeling the dams of the hydroelectric plants in question, care was taken to detail each dam's internal layers, as documented in the cross-sections of the As-Is documents. In **Figure 9**, the different layers discretized in the model of the Foz do Chapecó dam are visible. The asphalt core in its center was also modeled for this specific case.

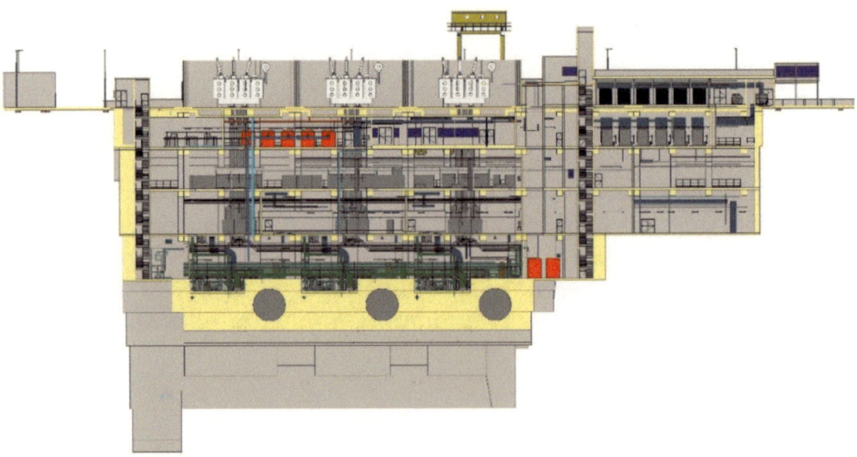

Figure 6.
Longitudinal section of Barra Grande's power house.

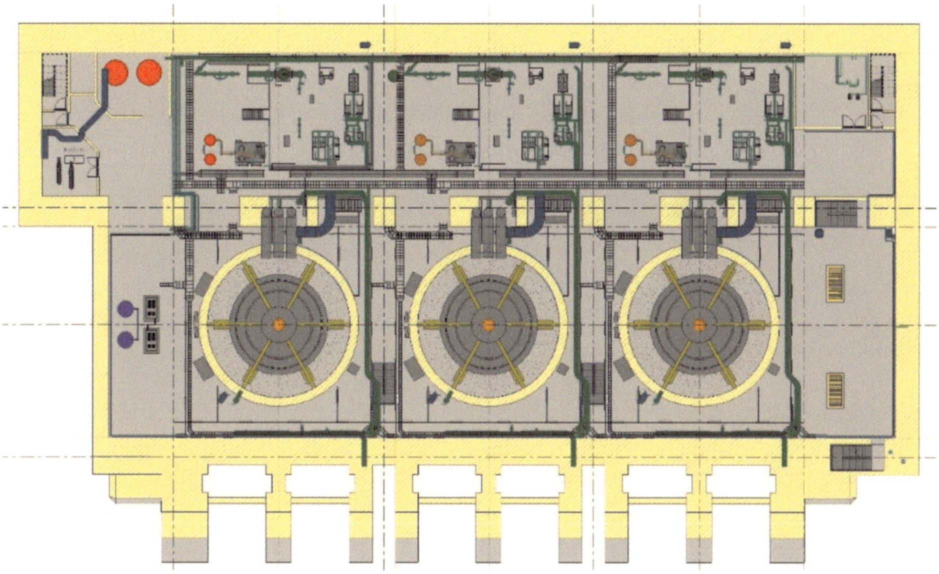

Figure 7.
Plan view of Barra Grande's powerhouse.

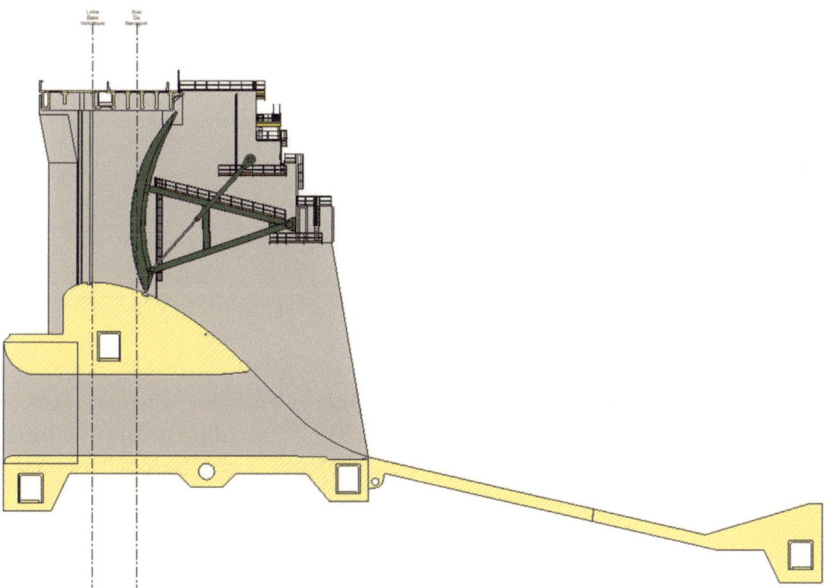

Figure 8.
Transversal section of Foz do Chapecó's spillway.

5.3 Eletromechanical

The modeling of electromechanical equipment is based on two essential premises. The first requires that the equipment be visually represented faithfully to its real counterpart. This involves analyzing characteristics such as material, color, and positioning. This premise aims to achieve a virtual representation that is as close to reality as possible, facilitating recognition in the virtual model.

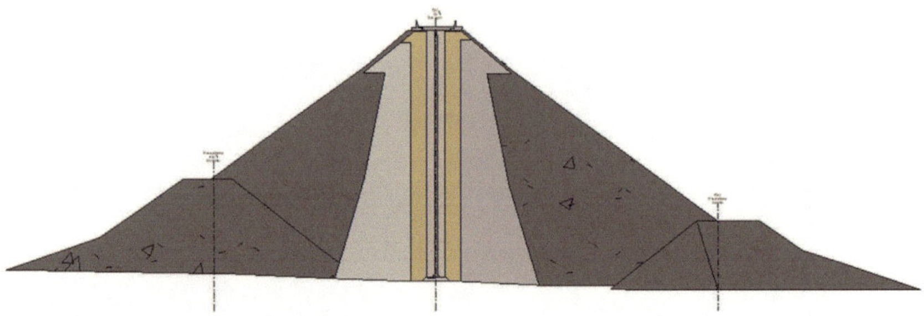

Figure 9.
Transversal section of Foz do Chapecó's dam.

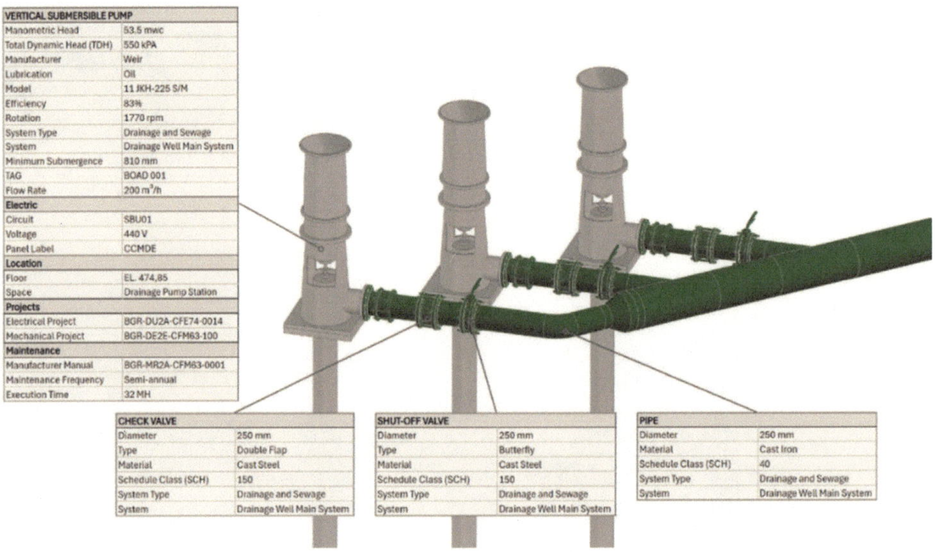

Figure 10.
Drainage well main system.

The second essential premise is that these elements must be parameterized to allow for the creation of various tools with the model. This approach enables the integration of information from different categories, such as location, mechanical, electrical, operational, and documentation aspects, into a single environment.

This factor creates an ideal context for integrating with hydroelectric plant management tools, such as SCADA (Supervisory Control and Data Acquisition) systems or ERP (Enterprise Resource Planning), maximizing the model's efficiency and usefulness.

Figure 10 shows a system model based on the premises above. It illustrates a set of elements that compose the drainage well main system. The pump, the most relevant element of the system in operational terms, has data from various categories and sources of information. Gathering this information required consulting the technical team, analyzing document specifications, and reviewing the plants' projects.

Structured data like those illustrated in the image enable the application of standards such as COBie. Crucial equipment details like material, technical characteristics, maintenance history, and warranty are structured in a standardized and interoperable form. This data standardization facilitates information exchange among various

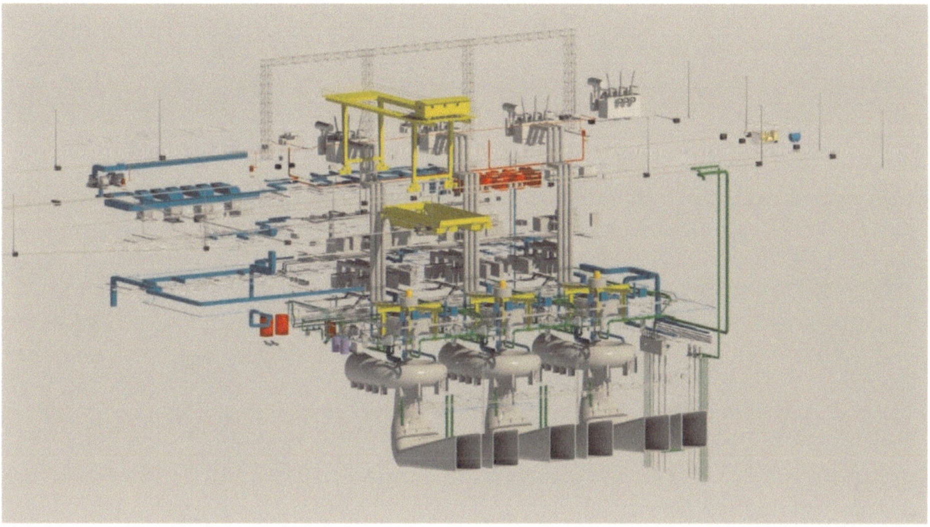

Figure 11.
Eletromechanical modeled systems—Barra Grande power house.

stakeholders, from engineers to facility managers, promoting optimal collaboration and decision-making.

Moreover, inputting data from projects and documents into the model makes it easier for plant maintenance and operation personnel to access relevant information. With data structured according to COBie, it is also feasible to conduct more precise performance analyses, identify potential faults, and optimize hydroelectric plant operations more efficiently, contributing to effective management and cost reduction.

Another important practice in modeling electromechanical equipment is dividing it into systems due to the complexity and large number of elements, as shown in **Figure 11**. As previously presented in this section, disciplines can be subdivided to enhance the categorization of elements. An application of this practice is creating color standards for each type of system and applying filters for these standards within the BIM software, which facilitates navigation and understanding of the model. In **Figures 11** and **12**, this feature is applied based on the pattern defined in **Table 2**.

5.4 Structural monitoring instrumentation

One of the project's main objectives was the creation of a Dam Safety tool. Consequently, instrument modeling was discussed and done separately from other disciplines, aiming to encompass all available information. The modeled instruments include 3D crack meters, drains, extensometers, flowmeters, inclinometers, pendulums, piezometers, prisms, settlement gauges, survey markers, tensiometers, thermometers, and total earth pressure cells. Basic parameters (TAG, georeferenced coordinates, and installation elevation) and specific parameters were assigned to all elements for all sensors. Additionally, specific parameters were defined for each type of sensor.

In **Figure 13**, a direct pendulum located at the 14 de Julho Plant is compared to the modeled one. In addition to the parameters mentioned above, which were addressed, there is also a concern for the visual fidelity of both. Therefore, besides the

Figure 12.
Power house scene using color filter.

System	Red	Green	Blue	Color
Low-voltage system	120	120	120	
Lighting system	120	120	120	
Fire protection system	255	0	0	
Service-compressed air system	0	128	255	
Lubricating and insulating oil system	196	89	17	
Ventilation system	0	80	128	
Cooling water system	0	128	0	
Potable water system	0	255	0	

Table 2.
Color standard adopted.

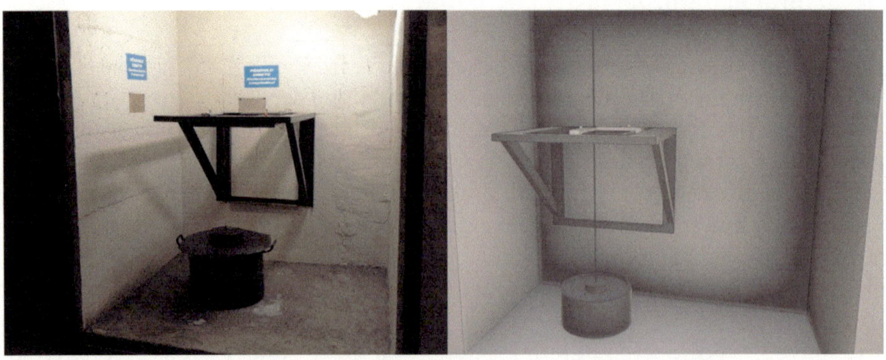

Figure 13.
Direct pendulum photo and virtual element.

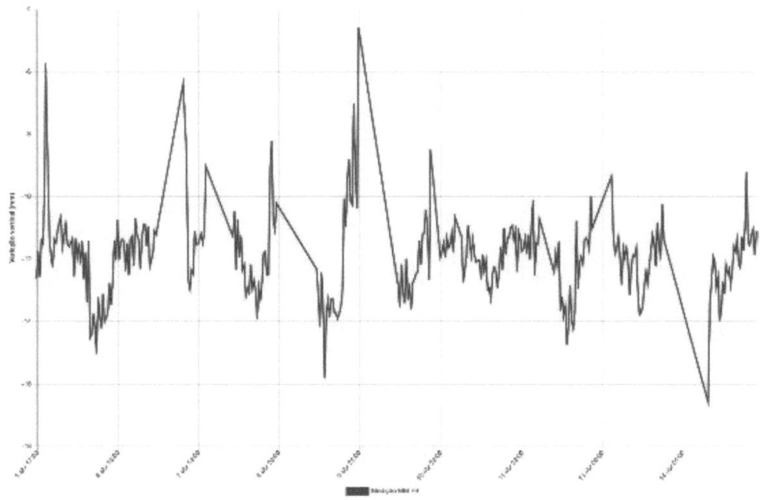

Figure 14.
Vertical displacement of a prism in Foz do Chapecó HPP main dam.

conventional access to reading data, these will be associated with an element within a georeferenced model. This allows for creating interactive visualizations and facilitates understanding for those not directly involved in the technical team.

Another factor considered in defining the modeling process was the subsequent integration between elements and their reading data. Both the available history and future measurements were associated with the equipment, making the safety management tool more robust. This tool, with its efficient use of the model and centralized structure management, provides reassurance about its benefits.

As an example of a result in DIGIDAM digital twin, **Figure 14** shows the vertical displacement of a prism in Foz do Chapecó HPP's main dam, which is monitored every 2 hours by a robotic total station (Leica TM 50i).

6. Conclusion

The use of BIM and VDC technologies in developing new projects and digitizing existing projects is undeniable. Each HPP project, or infrastructure project in general, is unique, with high complexity involving various disciplines and stakeholders, demanding specific planning for the project and using the developed models. This is likely a point that delayed the adoption of these practices. However, it has gained significant speed with the development of CDEs, allowing efficient management of many developed models and both structured and unstructured data.

Some CDEs interpret only Open BIM files as the IFC standard exported from the native software. Without diminishing the importance of the IFC standard for standardization, it has been observed that CDEs allowing direct connection with native software, such as ACC (Autodesk) and BCDE (Bentley), provide a considerable time gain in corrections and sheet maintenance.

For modeling already constructed works, planning the work to build a model that is scalable, both in terms of adding new structures and increasing the detailing of components, is essential. From a model standardization perspective, the case study followed, as much as possible, the guidelines of the IFC and COBie standards for both structured and unstructured data.

For complex structures like hydroelectric plants, the spatial division of structures to be modeled, then by floors, functional areas, and the main discipline groups within each structure, is essential for organizing models for CDE preparation.

In terms of the execution process, as these are relatively newly constructed plants with good and updated documentation, modeling started from existing drawings, with point clouds from laser scans used only selectively for installation adjustments. For older plants with a deficit of reliable documents, initial laser scanning and using available drawings as support will likely be preferable.

Considering that the primary focus of the model was the construction of a digital twin, more detailed work without immediate application, such as concrete reinforcement and construction stages, were disregarded. However, elements were created with project geometries and divided into expansion joints. In case of future need for more detailed work, each element can be treated separately and updated in the model. In a retrofitting scenario, this importance may change, and more structural detailing could be necessary.

Due to the benefits produced, digitizing the construction of a new project is practically mandatory. However, digitizing already constructed projects, both for providing digital twin information, improving asset and operation management, and retrofitting, proves extremely important and economically viable.

Acknowledgements

The authors would like to express their gratitude to the companies Foz do Chapecó, Baesa, Enercan, and Ceran for their technical and financial support through the Research and Development project (R&D) 00642-0123/2023—'BIM – Building Information Modeling Applied to Dam Safety,' funded by the ANEEL (National Electric Energy Agency) R&D program.

Conflict of interest

The authors declare no conflict of interest.

Author details

João Rodolfo Côrtes Pires[1], Paulo Sergio Franco Barbosa[2*], Pedro Nogueira Resende[1] and Ana Regina Mizrahy Cuperschmid[3]

1 Côrtes Consultoria, Campinas, SP, Brazil

2 Icptech: Innovation, Consulting, Projects and Technologies, Campinas, SP, Brazil

3 University of Campinas-Unicamp, Campinas, SP, Brazil

*Address all correspondence to: franco.barbosa@gmail.com

References

[1] Huang J, Bhat R. Thinking Inside of the Box: Using Autodesk BIM Technology for Hydropower Projects, Conference. Las Vegas, NV, USA: Autodesk University; 2016

[2] Adel M, Cheng Z, Lei Z. Integration of building information modeling (BIM) and virtual design and construction (VDC) with stick-built construction to implement digital construction: A Canadian general contractor's perspective. Buildings. 2022;**12**. DOI: 10.3390/buildings12091337

[3] Zielinski, Nguyen Ajslev J, Elisabeth Ejstrup Nimb I. Virtual design and construction for occupational safety and health purposes—A review on current gaps and directions for research and practice. Safety Science. 2022;**155**. DOI: 10.1016/j.ssci.2022.105876

[4] Lee ML, Cheah WT, Lau SH, et al. Evaluation of practicality of virtual design and construction (VDC) with 5D building information modelling (BIM) through a case study. In: IOP Conference Series: Materials Science and Engineering. Bristol, England: IOP Publishing; 2020. DOI: 10.1088/1757-899X/943/1/012058

[5] Kunz J, Fischer M. Virtual design and construction. Construction Management and Economics. 2020;**38**:355-363

[6] Garra RJ, Skripac B. IPD and VDC can lead to project success: Engineers can leverage the integrated project delivery and virtual design and construction processes. Consulting-Specifying Engineer. 2019;**56**:18-23

[7] Hassan H, Taib N, Rahman ZA. Virtual design and construction: A new communication in construction industry. In: ACM International Conference Proceeding Series. New York, USA: Association for Computing Machinery-ACI; 2018. pp. 110-113

[8] Said H, Reginato J. Planning and control framework for virtual design and construction: Case-based evidence from electrical construction. In: Construction Research Congress 2018: Construction Information Technology—Selected Papers from the Construction Research Congress 2018. Reston, Virginia, United States: American Society of Civil Engineers (ASCE); 2018. pp. 243-252

[9] Succar B. Building information modelling framework: A research and delivery foundation for industry stakeholders. Automation in Construction. 2009;**18**:357-375

[10] Wang M, Wang CC, Sepasgozar S, et al. A systematic review of digital technology adoption in off-site construction: Current status and future direction towards industry 4.0. Buildings. 2020;**10**:357-375. DOI: 10.3390/buildings10110204

[11] Al-Ashmori Y, Othman I, Rahmawati Y, et al. BIM benefits and its influence on the BIM implementation in Malaysia. Ain Shams Engineering Journal. 2020;**11**:1013-1019. DOI: 10.1016/j.asej.2020.02.002

[12] Toan NQ, Tâm N, Diep TN, et al. Adoption of building information modeling in the construction project life cycle: Benefits for stakeholders. Architecture and Engineering. 2022;**7**(1): 56-71. DOI: 10.23968/2500-0055-2022-7-1-56-71

[13] Barlish K, Sullivan K. How to measure the benefits of BIM—A case study approach. Automation in Construction. 2012;**24**:149-159

[14] Pasha P, Rasdi MTBM. Restructuring information Management in Building Design and Construction Using BIM as a platform. Civil Engineering and Architecture. 2022;**10**:600-619

[15] Nsimbe A, Di J. The impact of building information modeling technology on cost Management of Civil Engineering Projects: A case study of the Mombasa port area development project. Buildings. 2024;**14**(4):1175. DOI: 10.3390/buildings14041175

[16] Lu Y, Gong P, Tang Y, et al. BIM-integrated construction safety risk assessment at the design stage of building projects. Automation in Construction. 2021;**124**:103553

[17] Kaashi S, Vilventhan A. Development of a building information modelling based decision-making framework for green retrofitting of existing buildings. Journal of Building Engineering. 2023;**80**:108128

[18] Valinejadshoubi M, Moselhi O, Bagchi A. Integrating BIM into sensor-based facilities management operations. Journal of Facilities Management. 2022; **20**:385-400

[19] Astaneh, Asl B, Dossick CS. Immersive VR versus BIM for AEC team collaboration in remote 3D coordination processes. Buildings. 2022; **12**:1548. DOI: 10.3390/buildings12 101548

[20] Sampaio AZ, Sequeira P, Gomes AM, et al. BIM methodology in structural design: A practical case of collaboration, coordination, and integration. Buildings. 2023;**13**(1):31. DOI: 10.3390/buildings13010031

[21] Gulliver JS, Arndt REA. Hydropower Engineering Handbook. Chicago, IL, USA: McGraw Hill; 1991

[22] Singer D. Using Augmented Reality in Hydropower Plant Management and Maintenance [Master thesis] Univerza v Ljubljani Fakulteta za Gradbenistvo in Geodezio Lujbljana, Solovenia. 2022

[23] Ye H, Shang D, Liang T, Yan J, Hua T, Rui J. Research on integration application of BIM technology in water conservancy and hydropower automation system 2061–2065. In: 3rd International Academic Exchange Conference on Science and Technology Innovation (IAECST), Guangzhou, China, 2021. IEEE; 2022. pp. 2061-2065. DOI: 10.1109/IAECST 54258.2021.9695854

[24] Agostini M, Corbetti C, Ogbonna D, Stark M. Hydro's Digital Generation Transforming for the Future of Hydropower. Dublin, Ireland: Accenture's Insights; 2020

[25] Beken E, Guzmán R, Kruse M, Novoa A, Segura S. Winning the Productivity Race in Hydropower- Time for Agile Operational Innovation and a Digital Asset Management Mindset. 2021. Available from: https://www.adlittle.com/en/insights/report/winning-productivity-race-hydropower [Accessed: May 6, 2024]

[26] Autodesk. Reality capture and 3D models help protect critical infrastructure, the future of making things. Available from:https://www.autodesk.com/customer-stories/bureau-of-reclamation [Accessed: May 14, 2024]

[27] Bentley. Success story -Jinsha River Longkaikou hydropower station project: First hydropower station in China to use BIM technology. Available from: https://blog.bentley.com/software/success-story-jinsha-river-longkaikou-hydropower-station-project [Accessed: May 12, 2024]

[28] Lidar. Comprehensive 3D digital design on a China hydropower project

saves USD 300 Million. 2018. Available from: https://lidarnews.com/articles/comprehensive-3d-digital-design-on-a-china-hydropower-project-saves-usd-300-million [Accessed: May 12, 2024]

[29] Smith HB, Hansen GA. Construction of a hydropower plant with zero drawings: a success story from Norway. 2018. Available from: https://medium.com/autodesk-university/construction-of-a-hydropower-plant-with-zero-drawings-a-success-story-from-norway-2a116b4223c4 [Accessed: May 9, 2024]

[30] Multiconsult. Building a hydropower station without any drawings. 2016. Available from: https://www.multiconsultgroup.com/building-a-hydropower-station-without-any-drawings/ [Accessed: May 5, 2024]

Chapter 3

Analyzing the Emergence of Novel Hydroelectric Energy Markets: A Study on Turkey's Role and Development

Ibrahim Halil Demirel

Abstract

This study analyzes the current state and future prospects of the hydroelectric energy sector in Turkey. Firstly, it examines the development and current status of Turkey's hydroelectric energy sector. Subsequently, it addresses how government policies and the regulatory framework have shaped the hydroenergy sector in Turkey. Furthermore, the study explores the trends, key players, and projects influencing the hydroelectric market in Turkey, discussing the challenges and opportunities faced by the sector. Finally, it outlines Turkey's goals and future strategies for hydroelectric energy. This study underscores the importance of hydro energy for Turkey's sustainable development.

Keywords: hydroelectric energy, market, renewable energy, sustainable development, Turkey

1. Introduction

Hydroelectric energy is a major but often overlooked contributor to low-carbon electricity generation worldwide. It is the only low-carbon electricity generation technology that can provide system flexibility on timescales ranging from seconds to several hours and can store energy cost-effectively for days or months. These characteristics make hydroelectricity indispensable for energy systems. The critical contribution of hydroelectricity to the clean energy transition is not only its ability to generate large amounts of renewable electricity but also its unique flexibility and storage capacities for electric systems. These attributes make hydroelectricity a natural facilitator for the integration of more wind and solar energy. As variable renewable energy sources like wind and solar become more widespread, the role of hydroelectricity in providing balancing and reliability becomes increasingly important. Thus, hydroelectricity supports the transition to sustainable energy systems by enhancing energy supply security and reducing carbon emissions.

In developing economies, hydroelectric energy holds significant potential as the world's largest source of clean and flexible electricity generation; yet, this potential remains largely untapped. Despite its economic viability, a substantial amount of hydroelectric potential worldwide is still awaiting development. This potential is particularly evident in developing economies where energy demand is rapidly increasing, and many people still lack access to reliable energy services. When developed sustainably, hydroelectric energy can provide substantial benefits to the energy systems of many countries. It can play a critical role in enhancing energy supply security and reducing carbon emissions, especially by facilitating the integration of renewable energy sources [1]. In this context, ensuring environmental sustainability and social engagement in the planning and implementation of hydroelectric energy projects is crucial for long-term success. Balancing the economic, environmental, and social dimensions of hydroelectric energy will enable the effective utilization of this potential.

Despite numerous governments committing to achieving net-zero emissions by 2050, these targets often lack the support of reliable policies to facilitate significant emissions reductions. A concrete step to address this deficiency is the integration of hydroelectric power directly into energy and climate policy agendas. As a low-carbon energy source, hydroelectric power presents a valuable opportunity for policymakers aiming for a clean and secure energy future. Integrating hydroelectric power into energy policies can play a crucial role in achieving emissions reduction targets by increasing the utilization of renewable energy sources [1]. Fully harnessing and sustainably developing the potential of hydroelectric power will enhance the flexibility of energy systems and make significant contributions to combating climate change.

Turkey's geographical features, characterized by high mountains [2] and abundant water resources, provide substantial hydroelectric potential. The mountainous terrain facilitates water flow, creating ideal conditions for hydroelectric plants. This potential offers a significant opportunity to diversify Turkey's energy supply and ensure a sustainable energy source. Hydroelectric power plays a key role in Turkey's electricity generation portfolio, representing a substantial portion of the country's total electricity production.

The hydroelectric sector in Turkey has experienced rapid growth, particularly since the 1980s. During this period, numerous hydroelectric plants were constructed, driven by state incentives and investments alongside increasing interest from the private sector. Since the 2000s, liberalization and privatization policies in the energy sector have enabled significant developments within the hydroelectric sector. However, the sector faces various challenges in Turkey, including environmental impacts, sustainable water resource management, local community involvement, and social effects. Additionally, the potential effects of climate change on hydroelectric capacity must be considered.

Currently, priorities in Turkey's hydroelectric sector include sustainability, environmental impact consideration, increased use of renewable energy sources, and technological innovations. The future of Turkey's hydroelectric energy sector will be shaped by policy regulations, technological advancements, and international collaborations.

This article examines the current state and future potential of Turkey's hydroelectric energy sector. It begins by exploring the development and current status of Turkey's hydroelectric sector. Subsequently, it analyzes government policies and the regulatory framework to discuss how the sector is shaped by existing regulations and incentives. Additionally, the article addresses the challenges and opportunities faced by the hydroelectric sector and evaluates their impacts on its future in Turkey. Finally, the article outlines Turkey's goals and future strategies in the hydroelectric energy domain, providing an overview for readers to understand the article's content and flow, aiming for an in-depth examination of the current state of Turkey's hydroelectric energy sector.

2. Current status of hydro energy in Turkey

Over the past decade, Turkey has significantly diversified its energy production sources. Renewable electricity generation has tripled, and the commissioning of the country's first nuclear power plant will further diversify its fuel mix. However, fossil fuels continue to drive Turkey's economy, with substantial imports of oil and gas persisting.

In 2019, Turkey's high share of hydroelectric power and recent growth in other renewable energy sources positioned it above the average among the International Energy Agency (IEA) member countries in terms of the share of fossil fuels in electricity generation [3].

Turkey's gross theoretical hydroelectric potential is 433 billion kWh/year, with a technically exploitable potential of 216 billion kWh/year and an economically feasible potential of 180 billion kWh/year. Currently, the developed potential stands at 160 billion kWh/year, to fully develop the economically feasible potential through new projects. Turkey's theoretically exploitable hydroelectric potential represents 2% of the global theoretical potential and 18% of Europe's potential. With this potential, Turkey is the second-largest country in Europe, after Russia, in terms of hydroelectric capacity [4].

Turkey's total electricity production tends to increase in line with economic growth, population growth, and structural changes in the economy. In 2023, Turkey's electricity consumption decreased by 0.2% compared to the previous year, amounting to 330.3 TWh, while electricity production decreased by 0.6% to 326.3 TWh. According to the results of the Turkey National Energy Plan, electricity consumption is expected to reach 380.2 TWh in 2025, 455.3 TWh in 2030, and 510.5 TWh in 2035. As of the end of March 2024, Turkey's installed capacity has reached 107,959 MW. The distribution of installed capacity by source as of the end of March 2024 is shown in **Figure 1** [5].

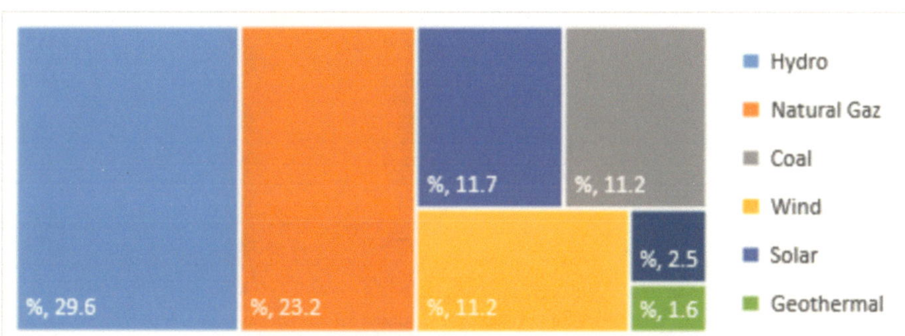

Figure 1.
Distribution of installed capacity by energy sources.

The change in hydroelectric installed capacity is presented in **Table 1**. From 2000 to 2010, the change was approximately 42%, while from 2010 to 2022, this figure increased by 99%. Over the past 5 years, the annual average increase in hydraulic installed capacity was around 4% in total. Ongoing hydroelectric power plant projects are still under development. Additionally, the installed capacities and project capacities of hydroelectric power plants (HEPP) are provided in **Table 2** [6].

Hydroelectric energy accounts for approximately 32% of Turkey's total electricity generation capacity, making it the leading source among renewable energy installations with a 60% share. As of December 2022, the number of HEPPs in Turkey reached 751 [8] (**Table 3**). Globally, Turkey ranks 9th in terms of hydroelectric installed capacity [9]. This position is attributed to the country's high elevations and abundant river resources. Despite high initial installation costs, factors such as the higher efficiency of hydroelectric energy compared to other renewable energy sources and various government incentives contribute to this success.

Years	Dam	N.lake and run of river	Total hydro	Increasing by year	Share in total %
2000	10,501.4	673.8	11,175.2	—	41.0
2001	10,959.4	713.5	11,672.9	4.5	41.2
2002	11,469.4	771.5	12,240.9	4.9	38.4
2003	11,752.4	826.3	12,578.7	2.8	35.3
2004	11,752.4	893.0	12,645.4	0.5	34.3
2005	11,967.4	938.6	12,906.1	2.1	33.2
2006	11,966.9	1095.8	13,062.7	1.2	32.2
2007	12,262.0	1132.9	13,394.9	2.5	32.8
2008	12,422.8	1405.9	13,828.7	3.2	33.1
2009	12,681.7	1871.7	14,553.3	5.2	32.5
2010	13,067.1	2764.2	15,831.2	8.8	32.0
2011	13,529.3	3607.7	17,137.1	8.2	32.4
2012	14,744.6	4864.8	19,609.4	14.4	34.4
2013	16,142.5	6146.6	22,289.0	13.7	34.8
2014	16,606.9	7036.3	23,643.2	6.1	34.0
2015	19,077.2	6790.6	25,867.8	9.4	35.4
2016	19,558.6	7122.5	26,681.1	3.1	34.0
2017	19,776.0	7497.1	27,273.1	2.2	32.0
2018	20,536.1	7755.3	28,291.4	3.7	31.9
2019	20,642.5	7860.5	28,503.0	0.7	31.2
2020	22,925.0	8058.9	30,983.9	8.7	32.3
2021	23,280.4	8212.2	31,492.6	1.6	31.5
2022	23,275.2	8296.3	31,571.5	0.3	30.4

Table 1.
Increase in renewables by year (MW) in Turkey, adapted from Ref. [7].

Situation	Power (MW)	%
In operation	32.180	72.9
Under construction	687	1.6
Production licence received	1.187	2.7
Associate's degree received	2.902	6.6
In project phase	7.202	16.3
Total	44.159	100

Table 2.
Hydroelectric installed power and project capacities in Turkey [6].

Plants	Run of River	Reservoir
EUAS plant	7	37
Power plants with transferred operating rights	69	24
Free generation company power plants	515	79
Build operates transfer power plants	1	1
Unlicensed power plants	18	0
Total	610	141

Table 3.
Project status and number of HEPP in Turkey.

3. Government policies and regulatory framework

The existing regulatory framework governing the hydroelectric energy sector in Turkey determines how activities within the sector are managed legally and institutionally. This framework encompasses licensing processes, environmental regulations, tariff policies, incentives, and other significant regulations within the sector. These regulations aim to support the sustainability and efficiency of sector activities by ensuring the consideration of environmental impacts and social factors in the planning, construction, operation, and maintenance of hydroelectric projects.

Electricity market reform in Turkey began with the enactment of the Electricity Market Law (EML) in March 2001, which aimed to increase competition and promote sustainable growth. The primary institutions regulating the hydroelectric energy sector include the Ministry of Energy and Natural Resources (MENR) and the Energy Market Regulatory Authority (EMRA). While MENR shapes energy policies, EMRA regulates and oversees the energy market.

In Turkey, energy flow typically involves energy produced by private companies or the Electricity Generation Company (EGC) being transmitted by the Turkish Electricity Transmission Company (TETC) to distribution companies and, ultimately, to consumers. Private producers can use distribution lines to provide energy directly to customers. Transmission lines belong to TETC, and distribution companies have the authority to operate these lines. TETC is responsible for electricity transmission, as well as system balancing and settlement. These tasks are carried out by two entities: the National Load Dispatch Center and the Market Financial Settlement Center.

Turkey's wholesale electricity market consists of three main markets: the Day-Ahead Market (DAM), the Intraday Market (IDM), and the Balancing Power Market (BPM). DAM and IDM are operated by the Energy Exchange Istanbul (EEI), while BPM is regulated by the grid operator TETC [10].

One of the other important regulatory frameworks is the environmental impact assessment (EIA) Regulation, which mandates the assessment of the environmental impacts of hydroelectric projects. Projects must undergo a comprehensive environmental impact assessment before construction, aiming to ensure environmental sustainability and local community participation. Another regulatory framework is the water usage rights and management framework, which governs the regulations related to water usage rights and management for hydroelectric projects. The General Directorate of State Hydraulic Works (SHW) is authorized for the management of water resources and the allocation of water to hydroelectric projects, with hydroelectric plants operating under water usage rights obtained from SHW.

The primary policies affecting HEPP in Turkey include:

a. Renewable Energy Resources Support Mechanism (YEKDEM): YEKDEM is a mechanism designed to support the utilization of renewable energy sources in Turkey. HEPPs are also supported under this mechanism. YEKDEM provides a fixed price guarantee and incentives for a certain period, increasing investor interest in hydroelectric projects. Additionally, it offers a tariff guarantee with additional incentives for the use of domestic equipment. Tariffs vary depending on the type of source and the percentage of domestic equipment used in the facility [11].

b. National Renewable Energy Action Plan (NEEAP): NEEAP aims to increase the share of renewable energy sources in Turkey's total energy production. Hydroelectric energy is a significant component of this plan, which seeks to increase the number and capacity of HEPPs.

c. Energy Efficiency Law and Related Regulations: The Energy Efficiency Law includes regulations aimed at enhancing energy savings and efficiency. HEPPs are considered energy-efficient projects under this law, and firms investing in this area are offered various tax reductions and incentives.

Other important government policies and incentives supporting the hydroelectric energy sector in Turkey include:

Financial Supports: The government offers various financial incentives for hydroelectric energy projects, including low-interest loans (as of the end of 2023, the Development Investment Bank has financed 6% of Turkey's renewable energy share [12]), tax reductions, grant programs, and investment incentive certificates. These supports aim to reduce the costs of hydroelectric projects and encourage investors.

Purchase Guarantees: The government provides purchase guarantees for hydroelectric energy production. These guarantees offer energy producers the assurance of selling electricity at a certain price, reducing the risk for investors. This facilitates the financing of hydroelectric projects and encourages investors.

Environmental Regulations and Licensing Process: The government implements various regulations and standards to assess and control the environmental impacts of hydroelectric projects. Additionally, there is a licensing process that specifies the procedures companies must follow to obtain a license for hydroelectric energy production.

These comprehensive policies and incentives are designed to promote the development of hydroelectric energy, ensuring its sustainable integration into Turkey's energy mix while addressing environmental and social considerations.

4. Market trends and drivers in Turkey

The IEA's May 2021 Net Zero by 2050 report indicates that to keep global temperature rise below 1.5 degrees Celsius, the world will need 2600 GW of hydroelectric capacity by mid-century. This implies that the capacity achieved over the past century will need to be doubled in the next 30 years [9, 13]. The report reveals that in 2020, the total increase in installed hydroelectric capacity was 21 GW, representing a growth rate of 1.6%, with an average growth rate of 1.8% over the past 5 years. Additionally, it is understood that approximately 1300 GW of new hydroelectric capacity will be needed by 2050 to maintain global warming below 1.5 degrees Celsius [14]. In 2019, 50 countries added a total of 15.6 GW of hydroelectric capacity, while in 2022, 35 countries added a total of 34 GW. Although the number of countries adding capacity has decreased, the total new capacity increased, raising the global hydroelectric capacity to 1397 GW. In Europe, the countries contributing the most to this increase include Portugal (998 MW), Switzerland (900 MW), and Turkey (558 MW). Consequently, hydroelectric energy accounts for more than 15% of global electricity production, providing approximately 4400 TWh [15].

Turkey's efforts to increase production and accelerate energy efficiency, aimed at achieving its energy security objectives, are among the country's strategic priorities. The rapid economic growth and population increase over the past two decades have led to a significant rise in energy demand, thereby increasing reliance on imports of oil and natural gas.

Turkey prioritizes energy supply security as one of the fundamental pillars of its energy strategy. Within this strategy, the expansion of domestic oil and gas exploration and production activities, such as the discovery of the giant Sakarya gas field in the Black Sea, plays a crucial role. This discovery has given significant momentum to efforts aimed at enhancing energy supply security [16]. Additionally, the diversification of oil and gas supply sources and infrastructure is a key component of Turkey's energy strategy. This diversification aims to reduce Turkey's energy dependency, ensuring a more sustainable and secure energy future.

In the last decade, Turkey has undergone a significant diversification process in its energy mix. Notably, renewable energy sources, led by hydroelectric, solar, and wind power, have demonstrated impressive growth supported by a conducive policy environment. This growth not only enhances energy supply security but also supports environmental sustainability. Furthermore, the planned commissioning of Turkey's first nuclear power plant in 2023 is expected to further diversify the energy mix, reinforcing the country's commitment to low-carbon energy sources [16]. This diversification is seen as a crucial step in ensuring Turkey's energy security and combating climate change.

Recent studies indicate that evaluating and analyzing Turkey's hydro resources is crucial for their potential use in hydrogen production [17]. As a green energy source, hydrogen offers a unique potential to reduce environmental impacts. The existing hydroenergy potential in various regions of Turkey is considered promising for hydrogen production and the hydrogen economy. This underscores the importance of hydroenergy and highlights that hydrogen could play a critical role in achieving Turkey's sustainable energy goals.

5. Key players and projects in Turkey

Turkey's continuous hydroelectric expansion constitutes a significant portion of growth in Europe, while opportunities for electricity exports in North America encourage the development of Canada's untapped potential. Outside these two markets, the main drivers of growth are renewable energy targets and the increasing need for system flexibility. However, environmental regulations and permitting challenges are key barriers to hydroelectric expansion. Consequently, developments are often directed toward projects where permitting is simpler, such as facility expansions and upgrades, or projects utilizing non-powered dams and existing reservoirs. Fleet modernization is emerging as a significant development activity in Europe and North America. Conversely, North America, Europe, and Eurasia will account for only 13% of global reservoir expansion over the next decade, as the number of economically viable sites decreases and social acceptance and permitting issues create challenges. Canada and Turkey are leading new site expansions in these regions, along with smaller capacity additions from facility renewals in other countries [1]. Notably, Turkey has yet to take concrete steps toward pumped storage hydroelectric (PSH) plants.

In this context, Turkey's hydroelectric energy sector holds significant growth potential. However, factors such as environmental regulations and permitting processes need to be carefully managed. Modernizing hydroelectric plants and increasing the efficiency of existing facilities will be crucial strategies to achieve renewable energy targets. Additionally, investments in PSH plants could enhance Turkey's energy supply security and facilitate the integration of renewable energy systems. These developments will contribute to the advancement of Turkey's energy strategies with a focus on sustainability and innovation.

In contrast, the private sector will be responsible for over 65% of global deployment in hydroelectric projects of 10 MW or smaller over the next decade. European countries currently hold the largest share of privately owned capacity. Between 2011 and 2020, over 80% of small hydroelectric plants in Europe were built by private firms, and approximately 70% of PSH are privately owned. The high share of private ownership in Europe will continue throughout the forecast period, with private developers contributing to about 70% of capacity additions, particularly in Turkey. Turkey has recently opened the way for private sector participation and increased the effectiveness of private investors in the hydroelectric energy sector. Specifically, Turkey has implemented a series of policies and regulations that enhance the role of the private sector in hydroelectric energy. In this context, mechanisms promoting the use of renewable energy sources and financial supports for the private sector have been provided. The increase in private sector investments in Turkey has contributed to the rapid expansion of hydroelectric capacity and increased diversity in energy production. Consequently, Turkey's energy strategies support energy supply security and sustainable development by encouraging private sector participation.

These developments enable Turkey to more effectively utilize its hydroelectric energy potential while also supporting its environmental sustainability goals. The strong involvement of the private sector accelerates the implementation of innovative technologies and efficient energy management systems, thereby enhancing Turkey's competitiveness in the energy sector. In this context, the continued and supported investments by the private sector are of critical importance for Turkey's energy future.

Global cumulative hydroelectric capacity is projected to rise from approximately 1330 GW in 2020 to just over 1555 GW by 2030, representing a 17% increase, or approximately 230 GW of additional capacity. However, the total capacity increase related to new hydroelectric turbine installations will be larger, reaching approximately 383 GW. This capacity increase will be divided between greenfield projects (new power plants) and activities at existing or abandoned sites (replacing or upgrading turbines or adding new turbines to existing facilities) [1]. This growth trend underscores the potential of hydroelectric power to enhance energy security and support sustainable development globally. The modernization and capacity expansion of existing facilities, alongside new projects, improve energy efficiency and minimize environmental impacts. Specifically, upgrading existing infrastructure simplifies environmental permitting processes and increases social acceptance.

In the early 2000s, debates continued in Turkey regarding the underutilization of hydroelectric potential. This situation led to the enactment of legal regulations emphasizing the evaluation of hydraulic energy under free market principles. Particularly, the law related to "Renewable Energy," which includes purchase guarantees and various support mechanisms, has garnered more interest from the private sector in this area. These laws and incentives facilitated the financing of hydraulic energy projects by reducing investor risk and increasing private sector participation in hydroelectric projects. As a result, there was a rapid increase in the number of both large and small-scale hydroelectric plants during this period. This increase significantly contributed to enhancing Turkey's renewable energy capacity and ensuring energy security. Especially, the purchase guarantees enhanced the economic feasibility of hydroelectric projects, making investments in this area significantly more attractive. Various support mechanisms also promoted the use of domestic equipment, contributing to local industry and supporting domestic production in the energy sector. Thus, energy supply security was increased, and the utilization rate of renewable energy sources was elevated.

The Black Sea Region is one of Turkey's richest areas in terms of hydroelectric potential. Particularly, the mountainous terrains in the Eastern Black Sea, with their high-flow rivers, offer significant advantages for hydroelectric energy production. Many large-scale HEPPs are located in this region. A notable example includes the Artvin province, which, with the Deriner and Yusufeli dams, boasts a total installed capacity of 2363 MW and ranks among the world's significant hydroelectric facilities.

Construction of the Yusufeli Dam began in 2013, and it was commissioned in November 2022, reaching full commercial operation by the summer of 2023. Standing at 275 meters, the Yusufeli Dam is the tallest in Turkey and ranks among the top 10 tallest dams globally. The dam has an installed capacity of 558 MW, featuring three 180 MW Francis turbine-generator units. With an annual energy production capacity of 1900 GWh, the Yusufeli Dam contributes approximately 269 million USD to the Turkish economy each year. Constructed at a cost of 1.8 billion USD, this facility supplies energy to 2.5 million households [18].

Turkey's largest hydroelectric project, the Southeastern Anatolia Project (GAP), focuses on irrigation and energy investments across the Euphrates and Tigris basins, aiming to promote economic and social development in the region. This project encompasses the provinces of Gaziantep, Adıyaman, Kilis, Şanlıurfa, Diyarbakır, Mardin, Siirt, Batman, and Şırnak, with a total population exceeding 9 million [19, 20]. As one of the largest investments of the Republic of Turkey, developed by SHW, GAP consists of 13 project packages, with seven in the Euphrates and six in the Tigris basins. The project includes the construction of various dams and HEPP with a total installed capacity of 7461 MW, expected to generate 27.3 billion kWh annually. Additionally, the project aims to irrigate 1.058 million hectares of agricultural land. As of 2020, 91% of energy projects and 55% of irrigation projects under the GAP Action Plan have been completed. Key installations such as the Keban, Karakaya, Atatürk, Birecik, and Karkamış dams and HEPP (totaling 6369 MW) on the Euphrates River, and the Batman, Kralkızı, Dicle, and Ilısu dams and HPPs (totaling 1612 MW) on the Tigris River and its tributaries have been completed and commenced electricity production. GAP is an integrated development project aimed at increasing the share of the region in the national income, alongside promoting socio-cultural development [21].

The Mediterranean Region also possesses hydroelectric potential, though it is more limited compared to the Eastern Black Sea and Eastern and Southeastern Anatolia regions. The rivers originating from the Taurus Mountains play a significant role in harnessing this potential. HEPP located in Antalya, Mersin, and Adana contributes to the region's energy production. These plants not only meet local energy needs but also support national energy supply security.

In summary, Turkey is developing various strategies and projects to reduce its dependency on foreign energy sources. Among these strategies, increasing the installed capacity based on renewable energy sources each year holds significant importance. Turkey's investments in renewable energy aim to enhance energy diversity, ensure energy security, and support environmental sustainability. In this context, hydroelectric energy stands out as a crucial component of Turkey's energy portfolio.

6. Challenges, opportunities, and government targets in Turkey

Europe's hydroelectric capacity is projected to increase by 8% between 2021 and 2030, equivalent to 18 GW. Nearly all of this growth is expected to come from new reservoir projects in Turkey, which will account for over 70% of new reservoir installations in Europe. The primary driver of this growth in Turkey is the country's hydroelectric targets, supported by guaranteed agreements for both large and small-scale hydroelectric projects. Specifically, private investments in projects ranging from 100 to 500 MW are expected to contribute to more than 70% of this increase. Turkey plans to install its first PSH units by 2030 to meet the rising share of carbon-free energy production and to enhance flexibility in the energy system. These units will not only bolster energy security but also facilitate the integration of renewable energy sources and significantly improve the flexibility of the energy system [1].

Outside of Turkey, the difficulty of obtaining permits for new hydroelectric facilities and the limited economic potential mean that new reservoir projects in Europe will likely be confined to Portugal and Croatia. The majority of the additional reservoir capacity in Europe will come from the expansion or refurbishment of existing aging facilities in countries such as Norway, Romania, Switzerland, Austria, and Poland. Modernization efforts in these countries aim to enhance the efficiency and sustainability of existing hydroelectric capacity. Furthermore, leading countries' global net hydroelectric capacity increases are detailed in **Figure 2** [22].

Nearly a quarter of global hydroelectric investments from 2021 to 2030 will be allocated to the modernization of existing hydroelectric fleets through turbine replacements, upgrades, and capacity increases. These investments aim to extend the lifespan of facilities, restore their initial performance levels, and enhance operational flexibility. The majority of global spending on modernization projects (75%) will be concentrated in North America, Europe, and Eurasia, where a significant portion of the world's aging hydroelectric capacity is located. By 2030, almost half of the hydroelectric fleets in these regions will be over 55 years old, necessitating major initial electromechanical investments to improve or maintain performance. Therefore, a substantial portion of these investments will be used to maintain the availability of facilities, restore performance, and adapt to operational conditions that may have changed since their original installation. Modernizing aging facilities will increase energy efficiency and support energy security.

Turkey also emphasizes the modernization of existing hydroelectric facilities in addition to its goals of increasing hydroelectric capacity. Many of Turkey's HEPPs were built in the second half of the 20th century and are now in need of modernization. In this context, SHW and EMRA are carrying out various projects for the rehabilitation and modernization of existing HEPPs. Within the scope of modernization projects, outdated turbines and generators are being renewed, automation systems are being updated, and the efficiency of facilities is being increased. Additionally,

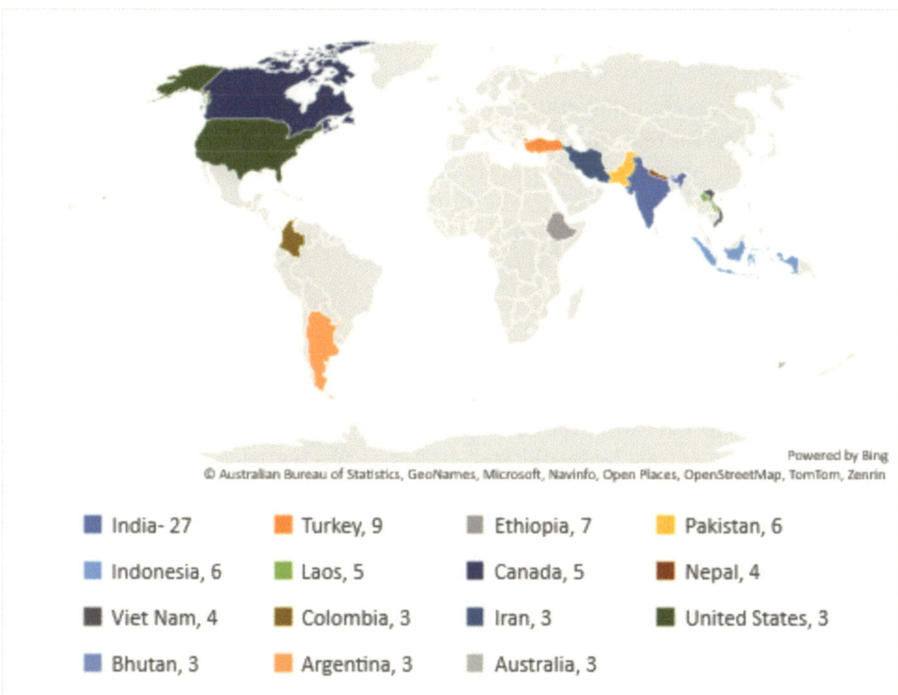

Figure 2.
Hydropower capacity (GW) additions by leading countries, 2021–2030.

innovative solutions are being implemented to enhance the flexibility of HEPP for the integration of renewable energy sources and to increase energy supply security. Turkey's modernization efforts not only increase energy production capacity but also contribute to long-term energy security by extending the operational life of hydroelectric facilities.

The second-largest source of growth in Europe is the expansion of PSH to provide the necessary system flexibility to integrate increasing shares of variable renewable energy. This expansion accounts for more than one-third of Europe's net hydroelectric growth. Adding pumping capabilities to existing infrastructure is a significant trend in PSH development in Europe, constituting one-fifth of PSH expansion. Such plants are less costly and easier to permit due to pre-existing infrastructure, such as natural reservoirs and reservoir HEPP being converted to mixed-use facilities. While Turkey is expected to install its first PSH units by 2030 to meet the growing shares of carbon-free production, it aims to reach a total of 35.1 GW in HEPP by 2035 to meet its energy needs [23].

Turkey's renewable energy employment differs from global trends. The diversification of renewable energy technologies also leads to diversification in employment opportunities. According to a report by the consulting firm Ernst & Young, Turkey has the potential to create 110,000 new jobs in a short period with an additional 9.2 GW of renewable energy capacity. This situation is closely related to countries' renewable energy potential, technology level, and economic status. In Turkey, employment opportunities in HEPP are relatively high compared to other renewable energy sources (43%). This is due to the large number of employees in HEPPs, which in turn results from the challenges during installation and the necessity for various sectors to work together [24].

Finally, changes in rainfall and flow patterns in Europe have become significantly apparent in recent years due to the impact of climate change. During this period, northern regions generally become wetter, while southern regions tend to become drier overall. It is expected that climate change will deepen these trends and make seasonal differences more pronounced over time [25]. This situation may adversely affect hydroelectric production in regions such as Turkey, Italy, Portugal, and Spain, which are southern European countries, while the consequences for northern countries indicate a more complex scenario.

7. Conclusion(s)

Hydroelectric power will continue to be a leading source of global low-carbon electricity production in the next decade.

Turkey's average annual hydraulic installed capacity increase in the last 5 years has been around 4% in total. Existing hydroelectric power plant projects are still ongoing.

To achieve renewable energy targets, the modernization of HEPP and the increase in efficiency of existing facilities will be a critical strategy. Additionally, investments in PHS in Turkey can enhance energy security and facilitate the integration of renewable energy systems. These developments will contribute to Turkey's energy strategies advancing sustainability and innovation.

Government needs to attach serious importance to policy to address current major challenges. Factors such as the cost and time required to conduct environmental assessments and ensure compliance with regulations, which can undermine the credibility of a project, pose significant obstacles for many projects. Therefore, support aimed at reducing project costs (such as financing for feasibility studies) and shortening licensing periods will accelerate growth.

It is of great importance that all hydroelectric projects anticipate the challenges caused by climate change in hydroelectric operation to ensure that they do not suffer due to the negative environmental and social side effects, and to effectively plan regional hydroelectric development and adapt operation rules appropriately.

Conflict of interest

The author declares no conflict of interest.

Abbreviations

SHW	state hydraulic works
IEA	international energy agency
MENR	ministry of energy and natural resources
EMRA	energy market regulatory authority
TETC	Turkish electricity transmission corporation
PSH	pumped storage hydropower
HEPP	hydroelectric power plant

Author details

Ibrahim Halil Demirel
Kozluk Vocational High School, Batman University, Batman, Turkey

*Address all correspondence to: ibrahimhdemirel@gmail.com;
ihalil.demirel@batman.edu.tr

References

[1] IEA, International Energy Agency. Hydropower Special Market Report-Analysis and Forecast to 2030. [Internet]. 2021. Available from: https://www.iea.org [Accessed: March 14, 2024]

[2] Ozalp AY, Akinci H, Temucin S. Mint: Determining topographic and some physical characteristics of the land in Artvin city and investigating relationship between these characteristics with land cover. Artvin Coruh University Journal of Forestry Faculty. 2013;**14**(2):292-309. DOI: ofd.artvin.edu.tr/issue/2267/29912

[3] IEA, International Energy Agency. Electricity Market Report-January 2022. [Internet]. 2022. Available from: https://www.iea.org [Accessed: March 02, 2023]

[4] SHW, State Hydraulic Works. Strategic Plan 2024-2028. [Internet]. 2023. Available from: https://www.dsi.gov.tr [Accessed: May 10, 2024]

[5] MENR, Ministry of Energy and Natural Resources. Distribution of Installed Capacity According to Resources. [Internet]. 2024. Available from: https://www.enerji.gov.tr/infobank-energy-electricity [Accessed: May 15, 2024]

[6] Energy Atlası. Turkey Hydroelectric Power Plants. [Internet]. 2022. Available from: https://www.enerjiatlasi.com/hidroelektrik [Accessed: May 10, 2024]

[7] TETC, Turkish Electricity Transmission Corporation. Development of the Share of Installed Power of Domestic Energy Resources in Turkey's Total Installed Power over the Years (2000-2022). [Internet]. 2022. Available from: https://www.teias.gov.tr/turkiye-elektrik-uretim-iletim-istatistikleri [Accessed: May 03, 2024].

[8] TETC, Turkish Electricity Transmission Corporation. Installed Power Reports. [Internet]. 2022. Available from: https://www.teias.gov.tr/kurulu-guc-raporlari [Accessed: May 17, 2024]

[9] IHA, International Hydropower Association. 2021 Hydropower Status Report. [Internet]. 2021. Available from: https://www.hydropower.org [Accessed: April 14, 2024]

[10] Barbaros E, Aydin I, Celebioglu K. Mint: Feasibility of pumped storage hydropower with existing pricing policy in Turkey. Renewable and Sustainable Energy Reviews. 2021;**136**:110449. DOI: 10.1016/j.rser.2020.110449

[11] Selcuk O, Acar B, Dastan SA. Mint: System integration costs of wind and hydropower generations in Turkey. Renewable and Sustainable Energy Reviews. 2022;**156**:111982. DOI: 10.1016/j.rser.2021.111982

[12] Development Investment Bank of Turkey. 2023 Operating Report. [Internet]. 2023. Available from: https://www.kalkinma.com.tr/assets/uploads/pdf/kalkinma_2023_Faaliyet_Raporu.pdf [Accessed: May 19, 2024]

[13] IEA, International Energy Agency. Net Zero by 2050 a Roadmap for the Global Energy Sector. [Internet]. 2021. Available from: https://www.iea.org [Accessed: May 19, 2024]

[14] IEA, International Energy Agency. Turkey 2021: Energy Policy Review. [Internet]. 2021. Available from: https://www.iea.org [Accessed: January 14, 2022]

[15] IHA, International Hydropower Association. 2023 World Hydropower

Outlook. [Internet]. 2023. Available from: https://www.hydropower.org/publications/2023-hydropower-status-report [Accessed: May 16, 2024]

[16] IEA, International Energy Agency. Turkey 2021 Energy Policy Review. [Internet]. 2021. Available from: https://www.iea.org [Accessed: April 15, 2024]

[17] Karayel GK, Javani N, Dincer I. Mint: Hydropower for green hydrogen production in Turkey. Internatioal Journal of Hydrogen Energy. 2023;**48**:2280-22817. DOI: 10.1016/j.ijhydene.2022.04.084

[18] IHA, International Hydropower Association. Region Profiles. [Internet]. 2024. Available from: https://www.hydropower.org/region-profiles/europe [Accessed: May 13, 2024]

[19] Energy Atlas. Turkey Hydroelectric Energy Potential Atlas. [Internet]. 2022. Available from: https://www.enerjiatlasi.com/hes-haritasi/turkiye, [Accessed: May 10, 2024]

[20] TSI, Turkish Statistical Institute. [Internet]. 2021. Available from: https://www.data.tuik.gov.tr/Bulten/Index?p=Adrese-Dayali-Nufus-Kayit-Sistemi-Sonuclari-2020-37210 [Accessed: February 06, 2022]

[21] MIT, Ministry of Industry and Technology. [Internet]. 2022. Available from: http://www.gap.gov.tr/en/what-is-gap-page-1.html [Accessed: May 19, 2024].

[22] IEA, International Energy Agency. Statistics. [Internet]. 2024. Available from: https://www.iea.org/data-and-statistics/charts/global-net-hydropower-capacity-additions-by-leading-countries-2021-2030 [Accessed: May 16, 2024]

[23] MENR, Ministry of Energy and Natural Resources. Türkiye National Energy Plan. [Internet]. 2022. Available from: https://www.enerji.gov.tr/Media/Dizin/EIGM/tr/Raporlar/TUEP/T%C3%BCrkiye_National_Energy_Plan.pdf [Accessed: April 10, 2024]

[24] Agbulut U, Yıldız G, Bakır H, Polat F, Bicen Y, Ergün A, et al. Mint: Current practices, potentials, challenges, future opportunities, environmental and economic assumptions for Türkiye's clean and sustainable energy policy: A comprehensive assessment. Sustainable Energy Technologies and Assessments. 2023;**56**:103019. DOI: 10.1016/j.seta.2023.103019

[25] EEA, European Environment Agency. Adaptation Challenges and Opportunities for the European Energy System: Building a Climate Resilient Low-Carbon Energy System, EEA Report No 1/2019. [Internet]. 2019. Available from: https://www.eea.europa.eu/publications/adaptation-in-energy-system [Accessed: April 10, 2024]

Chapter 4

Hydroelectric Power Plants and Their Sustainability

Moses Jeremiah Barasa Kabeyi and Oludolapo Akanni Olanrewaju

Abstract

Hydropower is the oldest source of renewable mechanical power and largest source of renewable electricity globally accounting for close to 17% of global electricity generation although its share of contribution has been diminishing gradually. Hydropower generation is a mature, reliable, predictable, and socially and economically competitive with high efficiencies greater than 90% in terms of water to wire conversion. There are different types of hydropower plants based on different criteria like size, application, and use of storage. Based on size they can be o micro, mini, small, or large hydropower plants, while based on storage, they can be either dam and reservoir plants or run-of-river stations. Different hydraulic turbines including the Pelton, Francis, and Propeller turbine are used in hydropower generation. Hydroelectric power plants have environmental impacts that should be considered and mitigated before development. The benefits of hydropower include low unit cost, low emissions, ease of control, and storage. However, hydro has challenges like large-land requirements for storage, soil erosion, ecosystem disturbance, and seasonal variability. It is technically feasible to combine hydropower with wind, solar, and other sources to develop a steady and affordable grid electricity mix. Hydropower plants with storage have additional benefits like flood control, water supply for industrial and domestic use, and power production for a sustainable grid in terms of flexibility and cost of power. Hydropower generally has higher initial investment cost but their long life span and lower operation and maintenance cost make hydropower competitive and sustainable option in green power generation.

Keywords: global warming, renewable sources, wind energy, climate change, hydropower generation

1. Introduction

Hydropower is the energy derived from falling or fast-running water that is converted to other useful forms like mechanical and electrical power. Hydropower can technically be defined as the rate at which hydraulic energy is extracted from moving or falling water resulting from velocity, position, or both. During generation, the change in angular momentum of falling water, the pressure, or both on the turbine blade surfaces creates a force on the turbine runner, which produces a rotary motion that can be exploited for electricity production [1]. In hydropower plants, the energy in moving water is converted to mechanical power by a hydraulic turbine, which turns a generator to produce electricity [2, 3]. Conventional hydropower plants have a reservoir developed behind a dam to supply water to the hydraulic turbine for the generation of a highly flexible, dispatchable electricity supply. Hydropower can be combined with wind, solar, and other sources to supply reliable, steady, and affordable grid electricity. Hydropower can also be exploited from run-of-the-river resources, which have less environmental impact but with overreliance on a steady supply of rainwater whose supply is unsteady and unpredictable. Apart from power generation, reservoirs can control floods, supply water, and power from stored water even during drought. Hydroelectric power plants are useful for grid electricity sustainability, particularly during peak hours when plants that generate flexible and cheaper electricity are in high demand [4].

In the last few decades, there has been a construction boom in hydro dams with more than 3700 major dams under development, each with an installed capacity of more than 1 MW, with most of the development taking place in emerging economies. With the installation of new dams worldwide the global hydroelectricity capacity will grow from about 1200 GW to about 1700 GW. Dams are important because they often contribute to the socioeconomic development of countries [5]. The installed capacity of hydropower was 1220 GW in 2022, which was an increase of 22.2 GW from 2021, with China leading in new capacity totaling 368 GW in 2022 [2].

Among the renewable energy sources, hydropower remains the largest source of renewable power by utilizing the energy in moving water for power generation. Hydropower is regarded as a clean energy source and can contribute to land reclamation and irrigation in times of drought, help stabilize the power grid and support socioeconomic activities like fishing, local tourism, and supply of domestic and municipal water [2, 6]. Over 159 countries globally have developed hydropower, which remains for generating renewable electricity at an efficiency of 90% or higher in terms of water-to-wire, making it one of the most efficient energy conversion systems [7].

This book chapter presents hydropower as a renewable energy resource, classification of hydropower plants, and hydraulic turbines used.

2. Main elements of hydroelectric power plants

A hydro power plant has a number of functional elements that enable it to perform its operation and control in power generation. The various parts are illustrated in **Figure 1**.

From **Figure 1**, it is noted that the main elements of a hydroelectric power plant are the reservoir, the dam, the forebay, the penstock, the surge tank, the turbine, the generator, the power house, the draft rube and the tailrace.

2.1 Dam

A dam is a structure used to create storage and elevation for the water surface and hence create a water head. The dam is the most expensive and important component of a hydropower plant, normally. Made of concrete, stone, and earth or rock fill. It is the topography or geography of the location that influences the type and configuration of a dam. Which may be a narrow canyon, or a brick dam, or an earth dam [8, 9].

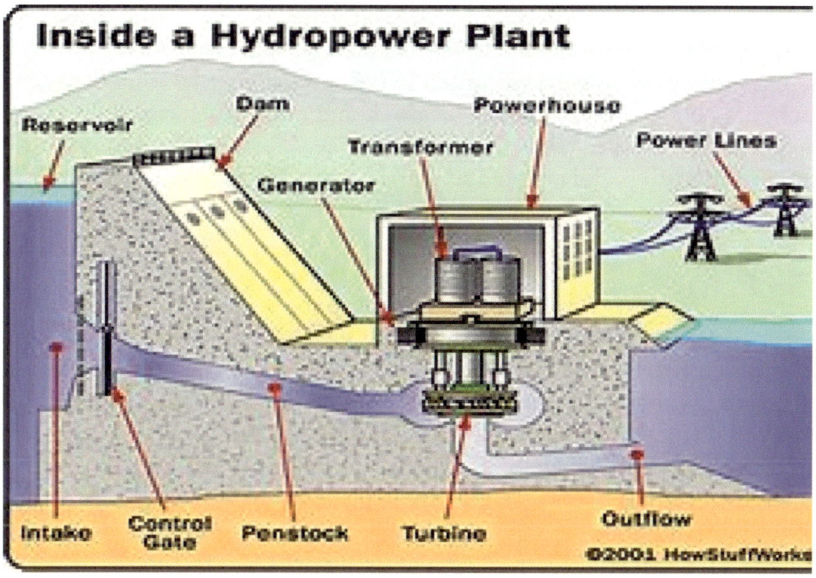

Figure 1.
Elements of a hydroelectric power plant.

2.2 Reservoir

The reservoir refers to a large water storage created by the use of a dam in most of the hydropower plants. The reservoir is created by creating a barrier across a natural water source like a river or lake, with water storage being created in the upstream side of the barrier. Reservoirs could be impounding reservoirs or balancing reservoirs, where the impounding reservoirs are large reservoirs with mechanically controlled gates for the opening and closing of gates to release the stored water based on needs. On the other hand, the balancing reservoirs are constructed downstream of the flow. And are mainly applied when there are water flow fluctuations [8, 9].

2.3 Forebay

The forebay is a basin area where the water is temporarily stored before entry into the intake structure of a hydropower plant. Chamber. The amount of water stored in the fore bay depends on the water demand by the plant as well as the supply pattern. The forebay also stores water when the load requirement is low or when a turbine rejects water, like in a trip or loss of load. Water leaving the reservoir via the dam is temporarily stored in the forebay which is a tank or a substructure that is connected penstock [8]. Therefore, a forebay refers to a temporary storage of a hydropower plant where water is held temporarily before entry to the penstock [9].

2.4 Intake structure

Intake structure receives the water from the forebay and directs it into the penstocks. The structure consists of components like the trash racks located at the entrance of the penstock to trap the debris in water. The selection of the appropriate intake structure is influenced by factors like local conditions. If left to enter the penstock, they may block or damage the wicket gates, turbine runners, nozzles, etc. The trash racks are usually made of steel rods. at spacings of 10–30 cm. The permissible water flow velocity is limited to 0.6–1.6 m/sec [8]. Other debris, the trash racks may block ice during cold weather, and under such conditions, the trash racks may be heated with electricity to melt the ice. Closing gates are also provided at the intake structure, while rakes and trolley arrangements may be used to clean the trash racks [8].

2.5 Penstock

The penstocks are pipelines or channels that deliver water from the forebay of the reservoir to the turbine in the power house. The water is primarily delivered under pressure to the inlet of a hydraulic turbine. The circular construction of the penstocks was purposefully built circularly to minimize the circumferential stress concentrations and enable even pressure distribution. The circular cross-section additionally reduces the possibility of scaling which is greater in square cross-section [8]. Penstocks are often made of mild steel due to high corrosion resistance and low friction loss and hence reduced head loss. Other materials used for penstocks are reinforced concrete. Penstocks can be exposed to the environment or buried based on geographical and environmental conditions [8].

The operational challenges include water hammer resulting from sudden closing or opening of penstock gates; hence they should be designed to withstand the water hammer effect. Measures include construction of heavy walls, short-length penstocks and surge tanks are installed on longer penstocks. A single penstock can be used for all turbines, or multiple penstocks with each turbine having its own penstock [9].

2.6 Surge tank

Surge tanks are water storage containers installed in the pathway of the penstock to serve as pressure neutralizers. They have inlet valves that bypass excess water from the channel and store it for temporary storage. In case of a drop in pressure in the channel, the outlet valve of the surge tank allows the stored water to exit the tank and join the flow in the channel to the turbine.

A surge chamber, or surge tank, is usually cylindrical and located at the top and as close as possible to the power house to control the penstock pressure caused by an increase or reduction in water demand due to demand fluctuation on the turbine generator. In case of load rejection, the water level in the surge tank rises to limit the pressure in the penstock. In case of sudden high demand in the power house, the surge tank releases water to increase the pressure in the penstock and help meet the increase in demand. The water level in the surge tank remains constant during instances of steady generation in the power house. The type of surge tank selected is influenced by the requirement of plant, and length of penstock among other factors [9].

2.7 Hydraulic turbines

Turbines are energy conversion devices that convert the energy in water to mechanical energy, which is used to spin the generator for power generation. The main elements of a hydraulic turbine are the runner, the guide blades, nozzles, the turbine shaft, the spiral casing, and the coupling to the generator. The kinetic energy or pressure of water, or both, acts as an inlet to the turbine to create a force component perpendicular to the central axis of the shaft that pushes the blades, which in turn creates angular velocity to the shaft [8]. In general, hydraulic turbines are either reaction turbines or impulse turbines based on the mechanism causing angular velocity. The impulse turbines are also called velocity turbines like the Pelton wheel turbines, while the reaction turbines are also called pressure turbines like the Francis and Kaplan turbines [9].

2.8 Draft tube

This is a long tube with a diverging cross-sectional area at the exit of a hydraulic turbine whose primary duty is to guide water exiting the turbine toward the tailrace. The design and shape of a draft tube which is diverging nature converts the kinetic energy of water leaving the turbine into pressure energy. The draft tube applies Bernoulli's theorem in its operation for the energy conversion in which pressure rises as the speed of flow reduces [8]. The negative head produced by the draft tube makes it possible to install a turbine above the ground without loss of the effective water head for power generation, which makes inspection and repairs of the hydraulic turbine easier [9].

2.9 Generator

The generator is a very important component coupled to the turbine shaft for conversion of the mechanical energy to electricity, which is the core function of a power plant. The operation of a generator is made possible by the magnets having a series of coils along with an armature. The generator armature rotation changes the magnetic flux, which induces a current flow in the coils. The current produced is then transferred to a step-up transformer and transmitted via electric cables [8].

2.10 Tailrace

The tailrace of a hydropower plant is a unit for the collection of water leaving the turbine on its way back to the original course of the river. If the power house is located far away from the stream, then a water channel is built to direct water into the stream. The design of the tailrace should allow smooth flow of water to the river without buildup which may damage the plant and lower turbine efficiency, induce cavitation, or damage the turbine blades and even cause power house flooding, etc. [8, 9].

2.11 Power house

The power house is a building that houses the turbine, generator, accessories, and the control room and hence protects the hydraulic and electrical equipment of the power plant, with the whole equipment being generally supported by the foundation or substructure that is laid for the power house [8]. For reaction turbines, some equipment or elements like draft tubes, scroll casing, etc., are fixed within the foundation during construction. For the superstructure, the generators are provided on the ground floor below which the vertical turbines are provided. The generator and horizontal turbines are generally laid on the first floor while the control room is either on the first floor or the mezzanine floor [9]. The general designs may, however, vary from one installation to another.

3. Classification of hydropower plants

Hydropower plants are broadly classified as conventional and pumped storage station power plants. The conventional hydropower stations account for the largest share of hydroelectric generation, while pumped storage accounts for the largest share of energy storage globally. Generally, a hydropower plant is where electricity is produced, having a dam for water flow control and a reservoir for water storage [10]. Hydropower, also called hydraulic power, hydrokinetic power, hydroelectricity, or simply water power, refers to energy that is derived from moving or falling water. This energy in moving water can be harnessed for useful applications like power generation or mechanical applications like milling. The first applications of hydropower for thousands of years were irrigation and the operation of various mechanical devices, like the watermills, sawmills, textile mills, dock cranes, and domestic elevators, among others. The latest applications of hydropower, which has taken over a century, are power generation in hydroelectric power plants mainly in centralized generation [11].

3.1 Size of hydropower plants

Hydroelectric power plants vary in size and range from small and micro small systems for a single home or village to huge central stations supplying thousands of MWe, supplying power to many consumers [12]. Hydropower plants can also be classified, based on their design or installed capacity, as large-scale, medium, small, mini, and micro plants, with mini, micro, and pico hydropower plants coming under the subcategory of small hydropower plants.

The small and even "micro" plants can be owned and operated by individuals for their own electricity needs or supply to the public [2]. The Department of Energy defines small hydropower plants as generally varying between 100 kilowatts and 10 MW [2]. Micro-hydropower has potential energy recovery sites that can capture excess energy within water infrastructure for the production of between 5 and 100 kW of power [13]. The micro-hydro plants are small, with a capacity of up to 100 kilowatts to supply single homes, farms, ranches, or villages [2, 14]. The various types of hydropower plants based on size are presented in **Table 1**.

From **Table 1**, it is noted that hydropower plants can be classified as pico, micro, mini, small, medium, and large based on the installed capacity, and sizes vary from as low as less than 5 kWe classified as pico hydro to greater than 30,000 kWe classified as large hydro.

	Category/size	Power range (kWe)
1	Pico	< 5kWe
2	Micro	+5kWe–100 kWe
3	Mini	+100 kWe–1000 kWe
4	Small	+1000 kWe–3000 kWe
5	Medium	+ 3000 kWe–30,000 kWe
6	Large	+30,000 kWe

Table 1.
Different sizes of hydropower plants [2].

There is a general belief that "small is beautiful" and hence small hydro plants have become popular since they are more environmentally friendly. As a result of this, small hydro schemes are more acceptable and even receive subsidies, while larger hydro schemes are not subsidized [15].

3.2 Classification based on head

The pressure head is determined by the level difference between the headwater and tailwater. On the basis of headwater, hydroelectric power plants may be classified into the following:

3.2.1 Low-head

The low-head installation generally consists of a dam across the stream to back up the river and create a fall of water through the turbines and remerging below the level of the dam. This is achieved by constructing a dam or barrage across the river to create a working or effective head. In such power plants, the dam and turbine are often very close, and thus there is no need for a surge tank. Commonly used turbines for low-head hydropower plants are the Kaplan, Francis, and propeller turbines. The discharge in such power plants is usually high; hence, bigger but shorter penstocks are often applied, while the structure for such plants is often extensive and expensive. Low-head plants often use low-speed generators compared to the high-head power plants [16]. The head of water for a low-*head* turbine is about 30 m or less.

3.2.2 Medium-head

In medium-head hydropower plants, the river water is usually tapped off to a forebay on one bank of the river and led to the turbines through penstocks. The forebay is installed at the beginning of the penstock and acts as a reservoir. The water is often delivered via open channels from the main reservoir to the forebay and then to the turbines through the penstock. The forebay acts as the surge tank, while the turbines used are often horizontal shaft Francis, propeller, or Kaplan turbines [16]. The water head for a *medium-head* turbine is between 30 and 300 m.

3.2.3 High-head

If high-head hydropower plants are located where a stream has a steep descent. The lateral valley can be dammed to form a reservoir for storage. A pressure tunnel is developed between the reservoirs and the power house to deliver water from the reservoir to the turbine, while a surge tank. is installed before the valve house to reduce the severity of the water hammer effect on the penstock just in case we have a sudden closure of the fixed gates of the water turbine or supply water to the turbine in case of a sudden rise in demand. Whereas there is no definite boundary between low, medium, and high heads, a water head above 300 meters is considered as high-head [16].

3.3 Based on design

On the basis of design, hydroelectric power stations can be broadly classified into run-of-river (ROR), storage (reservoir-based), pumped storage hydro (PSH), and in-stream (hydrokinetic) technologies [15].

3.3.1 Run-of-river power plant/diversion

The run-of-river *hydropower plants directly utilize the natural flow of the* stream by channeling water to the turbines and then directing it back into the natural flow, downstream of the plant [17, 18]. Therefore, the *runoff river hydropower* plants generate power mainly from the available flow in a river. Some systems make use of short-term storage (a pond) to facilitate adaptation to the consumption demand and pattern. However, the generation profile will generally follow the inflow profile of the river [2, 15].

The system uses a diversion, also referred to as a "run-of-river" facility, to direct part of the river through a canal and/or a penstock utilizing natural decline of the riverbed elevation to operate a hydro turbine for power generation. The water is directed through a penstock to turbines via regulated gate valves, which control the flow of water, hence power generation [19]. A typical run-of-river power plant has no water storage facility or may have a small one; hence, generation is directly influenced by the river flow characteristics, which may inhibit their dispatch capability in generation [2]. During normal water flow, the runoff river power plants are ideal for the supply of base load electricity.

The main limitation of a run-of-river plant without poundage during flooding or during low electricity demand is that excess water is wasted for lack of storage [20]. The firm capacity of the run-of-river *hydropower plants* is influenced by the minimum flow of the river. Whereas the water head keeps fluctuating, the plants can be designed to supply the base load, while generation is only possible when there is sufficient water flow in the river. At very low flows, water does not have to flow through the intake structure of the turbine. The run-of-river hydropower plants can be standalone systems, or can be grid connected [21].

3.3.2 Hydro with storage/impoundment/reservoir power plant

The reservoir *power plant uses water* from one or several streams/rivers collected *in a reservoir behind a natural or artificial dam* and then allowed to flow downstream through the turbines to produce power, then finally fed back into the river in a flow that is controlled in terms of flow and quantity to produce dispatchable power [17, 18].

These hydroelectric power plants having a reservoir store water when the flow is high, like during flooding, the rainy season, or low demand, for later use, when the flow is low or demand is high. A reservoir facilitates high flexibility necessary for the hydropower plant to adapt better to the varying demand profile [15]. Impoundment hydropower facilities are the most common type of hydroelectric power plants. They are typically large hydropower systems that have a dam for storage of water in a reservoir for release through a turbine, thus spinning it to activate a generator to produce electricity. Power generation in the impoundment hydroelectric power station is made in a manner that meets power demand and applications like flood control, fish recreation, and other environmental and water quality needs [19].

Elements of a hydro scheme with storage include the dam, reservoir, trash racks for stopping foreign matter, penstock, surge tank, and turbines. A surge tank is used to prevent sudden pressure rises in the penstock due to drops in turbine load as well as supply extra water to the turbine during sudden load rises. A controllable valve controls the flow of water through the hydraulic turbine for power generation, and then used water is discharged to the tailrace [22].

3.4 Classification according to type of load supplied

Hydraulic turbines can also be classified on the basis of the load supplied into the base load. Peak load and pumped storage plants are usually for peak load [16].

3.4.1 Base load plants

Base load power plants supply load on the base portion of the load curve and are usually larger capacity power plants. These power plants operate almost on constant load throughout at a higher load factor. The run-of-river plants without pondage and reservoir plants are often used as base load power plants. To qualify as base load power plants, they should have a low unit cost of energy generated [16].

3.4.2 Peak load plants

Peak load hydropower plants supply electricity for peak load, e.g., the power plants having poundage. For power plants with bigger poundage, a large portion of the load can be supplied by when required. The reservoir power plants are generally used and are often operated during the peak and often used to store water for dry seasons too [16].

3.4.3 Pumped storage hydro (PSH)

The first ever documented pumped storage hydro (PSH) was located in Italy and Switzerland in the 1890s, while in the US, the first PSH was developed and used in 1930 and accounts for about 93% of all utility-scale energy storage in the United States from at least 43 pumped hydro storage power plants with significant potential to increase the current capacity [23]. The Pumped-storage hydroelectricity (PSH), also referred to as pumped hydroelectric energy storage (PHES), is a hydroelectric power plant mainly used for load balancing. Pumped storage hydro is used to balance day-to-day power generation and load by pumping water to tall storage dams during off-peak hours using cheap and surplus base load power. This water storage is then used to run hydraulic turbines for peak power generation. The pumped storage hydro systems are equipped with a Pump as turbine (PAT), which is used for pumping as well as a turbine in the reverse direction. The PSH system is the best option for meeting fluctuating demand because of its ability for quickly ramp-up as well as ramp-down, economically based on differences between peak and off-peak wholesale charges for electricity [2, 23]. As a result of water pumping, PHS are net consumers of energy but useful for balancing and making profits on price differences for electricity [24]. Pumped storage hydro (PSH) can be a closed or open-loop type depending on the existence and connection to a natural source of water like a river. The open-loop PSH, is connected to a natural source of water like a river, while the closed-loop PSH system is not connected to an outside body of water or water body [23, 25].

The use of PSH facilitates the uptake of electricity from variable and or excess electricity from continuous base load plants like nuclear and coal for storage and use during the peak demand periods [24]. Pumped storage hydropower accounts for the grid-based storage, which in 2020 accounted for 95% of global energy of over 181 GW, operating at a round-trip energy efficiency of 70–87%. Locations for siting PSH are generally limited to hilly or mountainous regions, making PSH susceptible to social and ecological concerns [3, 24].

4. Hydraulic turbines

Hydraulic turbines are prime movers in hydropower plants that come in a variety of designs for different operational and environmental conditions. It is important to select the most suitable hydraulic turbine for a given specific application; hence, energy developers should understand the various designs, performance, and applications [26].

There is a number of hydroelectric power systems whose application is dependent on the environmental physical conditions of the site, water discharge, water head, and demand or capacity, which also influence the type of hydraulic turbine selected [16]. The hydraulic turbines are used to convert the energy in water into rotational mechanical energy, which can be transformed into electricity by the use of the generators coupled to the turbines [27]. They can be classified into two main types based on water pressure, i.e., reaction and impulse turbines [3, 16].

In the impulse turbines, water pressure remains constant while passing through the turbine rotor. When nozzles are used, energy in water is converted to kinetic energy as it flows through the nozzles, often in the form of jets that strike the bucket-shaped vanes located on a rotating shaft, thus converting the kinetic energy of the water to mechanical rotation of the impulse turbine [2, 27]. On the basis of the direction of flow, the turbines can be classified into axial flow, crossflow, and mixed flow turbines. The hydraulic turbines can also be classified based on the shape and orientation of the turbines, e.g., vertical turbines, bulb turbines, and Strallo turbines. The choice of the hydraulic turbine is influenced by the water height, velocity of flow, and the discharge at the site of the plant. Other considerations in turbine selection are cost, turbine efficiency, and depth of turbine installation [27, 28]. Modern hydro turbines vary in design, i.e., size, with "runner" diameters from as small as a third of a meter to some 20 times this. The turbines may also be reaction time or impact type, high-head or low-head turbines, etc. The various types of turbines are discussed in **Figure 2**.

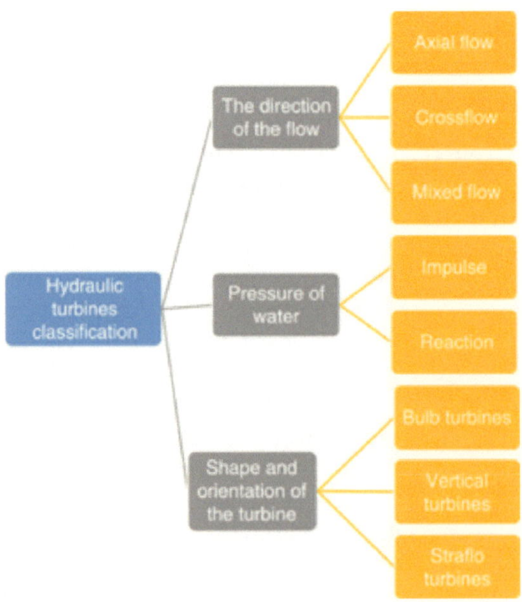

Figure 2.
Types of hydraulic turbines [3].

There are different types of hydraulic turbines that can be used in power generation. They can be classified on the basis of water pressure, the direction of flow, or the shape and orientation of the turbines. Hydraulic turbines generally fall into two main categories: namely, the impulse turbines and the reaction turbines.

The objective of any hydro turbine is the ability to generate electricity by the rotation of the shaft, hence the need to carefully select a turbine that is ideal for the energy in the water to and site characteristic [29]. In terms of the head, turbines are classified as

i. Low-head turbines (up to 40 m) e.g., the propeller and Kaplan Turbine.

ii. Medium-head turbines (40–100 m) include the Francis, pump as turbine, cross-flow and Pelton wheel.

iii. High-head turbines (>100 m) e.g., the Turgo and Pelton wheel [29].

4.1 Reaction turbine

A hydropower electric power plant is designed to utilize the available water and head for power generation using different types of hydraulic turbines, e.g., the Pelton and Francis turbines for high- and medium-head hydropower plants, while the Kaplan turbine is preferred for low-head and large-flow hydro systems [15]. Reaction turbines are a combination of the pressure forces and moving water using a runner installed in the water stream for water to flow over the blades, usually at sites having low water head but higher flow rates [28]. The Kaplan and Francis turbines are the two commonly used types of reaction turbines [2, 30]. The water pressure varies based on variations in the shape of the flow path shape, consisting of fixed guide vanes, rotating blades, and adjustable guide vanes [27].

In reaction turbines, water at high pressure enters the turbine tangentially to the runner via the guide vanes which are fixed, then directed to adjustable guide vanes at high tangential velocity components. As the runner rotates, momentum is swapped between the fluid and the runner, as water fills the casing and producing power, which is higher compared to the impulse turbine having similar diameter, head, and water flow rate [2, 27].

4.2 Impulse turbines

The history of impulse turbines is based on early designs by Ferdinand Veribest's impulse steam turbine-driven car in 1672 and Frenchmen Real and Pichon, who patented the compound impulse turbine of 1827. The name "impulse turbine" is derived from operation, relying on the force generated by the impact of water jets on the turbine blades. Impulse turbines convert the velocity of moving water in the form of jets or streams that hit buckets mounted on the runner without any suction, thus rotating it and discharge it at atmospheric pressure. The used water flows out through the bottom of the turbine housing. These turbines are ideal where there is a high head and low flow rates. The water tangentially strikes the blades within impulse turbines, hence the alternate name, the tangential turbine, and they are well suited for high head and low flow rates, with water at an elevated position creating increased pressure. The Pelton turbine is the most common type of impulse turbine, followed by the cross-flow turbine [28].

Impulse turbines convert the kinetic energy of a high-velocity fluid into mechanical energy, which drives a generator or other mechanical devices. The impulse turbines work by directing high-speed water jets onto a series of specially shaped buckets or blades where the force of the jets rotates the turbine blades and turbine shaft. The impulse turbines are preferably used in hydropower plants and other applications having high-speed flows of water that offer efficient and reliable power generation [2].

4.3 Propeller turbine

The propeller turbine has a runner with three to six fixed or adjustable blades and is operated as water constantly contacts all the blades. Other components of the propeller turbine are the runner, draft tube, scroll case, and wicket gates [28]. Propeller, or axial flow turbines operate by sweeping the blades through the entire area in which water enters, hence suitable for large water volumes of flow where the water head is a few meters. The efficiency of propeller turbines can be improved by varying the angle of the blades as demand changes [2].

4.4 Tubular turbine

Tubular or tube turbines are horizontal or slant-mounted turbines equipped with propeller runners, with the generators located outside of the water passageway. The turbines have fixed or variable pitch runners and with or without wicket gate assemblies. The performance characteristics of a tube turbine are similar to those of the propeller turbines. Tubular turbine efficiency is 1–2% higher than that of the vertical propeller turbine of the same size due to reduced change in direction of water passageway. The performance of a tube turbine having a variable pitch blade and without wicket gates is better than that of the fixed-blade propeller turbine but less than the performance of the Kaplan turbine. The flow of water through the tubular turbine is achieved by changing the pitch of the runner blades. A fixed blade is used in applications that do not need regulation of discharge and power output of the turbine.

Tube turbines without wicket gates are equipped with a shut-off valve that is automatically operated for provision of shut-off and start-up functions. They are connected either directly to the generator or through a speed increaser. Which facilitates the use of a higher speed generator, typically 750 or 1000 (1500) r/min. The challenge of using speed increasers is reducing the overall plant efficiency by about 1% for a single gear increaser and close to 2% for a double gear increaser and may not be ideal for units above 5 MW [30, 31].

Their tubular turbines make the best use of the strengths of a Kaplan runner without a spiral case and draft tube. There are different types of hydroelectric units equipped with a tubular turbine; e.g., a bulb unit has a unit generator housed in a bulb in the water passage, with the generator directly driven or indirectly via a gearbox. Other types are "Pit units." The pit units have a generator housed in a pit in the water passage and allow direct dismantling of the generator and the gearbox from above [2].

Technically, the tubular pumps are highly efficient, with rates of up to 90% or more, while the tubular hydro generator, which typically consists of magnets, copper wire coils, and a rotating shaft, is designed to convert the mechanical energy of a tubular water turbine to electricity, operating on the principles of electromagnetic induction. Tubular turbines can be used for small-scale, mini- to large-scale hydroelectric dams to produce clean and renewable energy, with no emissions or pollutants [30].

4.5 Bulb turbine

The Bulb turbine integrates the generator in a waterproof casing in the stream of water used in tidal and low-head river applications. The direct drive without gearing reduces transmission energy losses. The Bulb turbines are horizontal-axis turbines with propeller runners directly connected to the generator, which is enclosed in a watertight enclosure (bulb) located in the turbine water passageway. The turbine has got either fixed or variable pitch blades with a wicket gate mechanism or none [26].

The Bulb turbine consists of a turbine generator as a sealed unit placed directly in the stream of water. The Bulb turbine is also known as the "horizontal-axis-propeller" turbine, equipped with variable pitch blades. Bulb turbines have bulb-shaped casings containing generators and other auxiliary systems and also acts as nozzles for increasing the velocity of the flow of the water as it approaches the blades. A large bulb is installed at the center of the water pipe supporting the generator, rotating blades, and the wicket gate [27]. The Bulb turbine is both lubricated and cooled by the water [32].

In the enclosed, bulb-shaped, sealed structure, the generator and turbine are rigidly fixed. The arrangement applies to active pumps and power generation turbines. A reversible pumping generator is applied in overfilling the tidal basins at the high tide level and often to empty the tidal basins at the low tide level [32, 33].

On performance, a bulb turbine has a relatively higher efficiency of about 2% higher than the vertical unit and 1% more than the tube unit because of the straight water passageway. They have lower space requirements because of the compact design but require longer time during maintenance because of accessibility challenge; however, it may be greater than for either the vertical or the tube-type turbines [32].

4.6 Straflo turbine

The Straflo turbine incorporates a generator at the periphery of the turbine's runner, which facilitates a direct connection between the generator and the turbine to enhance the overall compactness and efficiency. This turbine is ideal for low-head hydroelectric plants. Straflo turbines, also called the straight flow turbines, are axial turbine-generators located outside the water delivery tube connected to the outer ring on the edge of the rotor blades. A sealing is installed to limit water leakage from the rotor blades to the generator [27, 32]. The Straflo turbine is an axial flow turbine that is a fully tubular machine operating at high specific speed used on high discharge and low head. In the Straflo, the generator is attached directly to the perimeter of the turbine. Straflo turbines comprise a turbine and generator assembled as a single unit and are best used in micro-hydropower, where water flows in pipes. The generator is mounted around the periphery of the runner. An efficient Straflow turbine design needs a detailed flow behavior inside the turbine space that largely depends on the runner geometry and distributor blades [34]. Straflo turbine is the ideal choice for the hydropower plants where water is delivered through a pipeline at a slope and is converted into mechanical energy in turbine [34]. The axial flow turbine blades are highly twisted, leading to a change in angular momentum that forces the rotor to rotate along with the generator shaft to generate electricity [34, 35].

In this turbine, the distributor, runner, and draft tube are designed for a fixed head and discharge arrangement. However, for most power plants, the turbine operates at off-design values in terms of the head and discharge. These changes in the discharge and the head change local flow parameters, like velocity and angles, which affect turbine performance, making it customary to assess flow and performance behavior of the turbine in detail for off-design operating conditions for efficient design [34].

4.7 Kaplan turbine

This turbine was developed in 1913 by Viktor Kaplan, an Austrian inventor [28]. The Kaplan turbines are axial flow turbines with variable blade angles to improve efficiency. The turbines have both adjustable blades and wicket gates that make them effective over a wider range of operating conditions [28]. The turbine operates based on the principle of operation of a boat propeller. Flowing water turns the propeller blades as it enters and exits axially relative to the propeller's axis of rotation. Optimum performance of the Kaplan turbine is made possible by adjusting the blades' angle of incidence. The turbine has optimum performance with *small height differences, as* well **as** *large variations in capacity,* i.e., 200 m^3 per second and above [17].

The Kaplan turbine is the preferred choice for applications on low-head installations ranging from 2 to 70 m and high water flow rates of 70–1000 m3/s and net power up to 200 MW [27].

4.8 Francis turbine

The Francis turbine is the most common type of hydraulic turbine used in medium- or large-scale power plants where the head is medium- to high-head, as high as 200 m, but it can also be used selectively for lower heads too of below 2 m. The Francis turbine was invented by a British-American engineer named James Francis in 1848. The process that developed the modern Francis runner design took place between 1848 and about 1920 and became known as the Francis turbine around 1920 [17].

The Francis turbine has a runner with generally nine or more fixed blades, with more blades in which water is introduced above the runner and then all around it before it falls through, forcing the blades to rotate. The draft tube, scroll case, and wicket gates are the main components of a Francis turbine are. The turbines work well in both vertical and horizontal axis orientations [28]. The Francis turbine is a radial-flow turbine in which water flows inwards toward the center. The turbines can achieve as high as 95% efficiency under optimum conditions in terms of speed and direction of the incoming water relative to the runner blades [17].

In the Francis turbine, water reaches the impeller through a spiral conduit, while an adjustable palette, which is attached to the stationary part, guides the flow of water to strike the impeller blades. The turbine is applied on *medium height differences*, from 10 to 400 meters, and from *2 to 100* m^3/s of water discharge/flow capacity [17].

The Francis turbine combines impulse and reaction actions and can effectively operate under a head between 40 and 600 m and mass flow of 2–800 m3/s and net output of up to 800 MW. The Francis turbine is flexible and can operate efficiently in wide range of operating conditions [27].

4.9 Kinetic turbine

Kinetic energy turbines, also called free-flow turbines, produce power by converting kinetic energy of water that is contained in flowing water instead of the potential energy from the water head of rivers, man-made channels, tidal waters, or ocean water currents. The kinetic energy turbines do not require a diversion of water through pipes, riverbeds, or man-made channels or canals, even though they can still be used. Therefore, the hydraulic turbines can be installed on natural structures with little or minimal civil works [28]. Hydrokinetic remains less developed technology and least used than the other three. However, hydrokinetic has the potential to extend the range of feasible hydropower development sites to rivers and canals for direct energy harvesting from flowing water, instead of using a hydraulic head created by dams or other control structures [15, 36].

The hydrokinetic, or kinetic turbines, constitute a class of renewable energy devices that convert the energy in moving water. They can also be used in ocean-energy structures, where they can be arranged into multi-unit arrays to extract energy from tidal and marine currents.

The hydrokinetic energy production avoids limitations of the traditional hydropower units, like high civil works costs and high potential energy head. Although hydrokinetic turbines are of the oldest hydropower, the systems have evolved from the waterwheel concept to the recent modernized pilot plants and experimentation, and currently there are several hydrokinetic designs for use in river, tidal, and wave energy systems [2, 37].

4.10 Pelton turbine

Pelton turbine is the most popular impulse turbines suitable for high water heads of 200–4000 m but low water flow rates of between 0.5 and 20 m^3/s and net power of up to 200 MW, at an efficiency of about 90% [27]. The Pelton is an impulse turbine invented by Lester Allan Pelton, an American inventor, in 1879. The Pelton turbine is ideal where the heads are above 250 meters operating at normal atmospheric pressure. The Pelton turbine is basically a wheel having a set of double cups or "buckets" mounted around the rim. It is operated as water impinges the buckets and passes round the curved bowls. One or more free jets that impeach on the buckets of a runner installed in an aerated space. Pelton turbines are suitable for applications with very high water heads but low discharge. The Pelton turbines do not need draft tubes because the runner is located above the maximum tail water to enable operation at atmospheric pressure [28].

When operating under optimum conditions, the Pelton turbine gives up almost all its kinetic energy. Power can be adjusted by varying the jet size or by deflecting the entire jet away from the wheel. Which alters the flow rate over the turbine [17].

The principle of operation of the Pelton turbine follows the classic paddle wheel of old watermills but is redesigned for higher efficiency. The jet of water emerging from a nozzle strikes the impeller's spoon-shaped blades, causing the wheel to rotate. For higher velocity to be achieved, Pelton turbines are applied in large height differences, from 300 up to 1400 meters, and lower discharge of less than 50 m^3/s of water capacity [17].

4.11 Turgo turbines

This is an impulse-type turbine used for medium-head applications and achieves efficiencies of about 87–90% operating at net heads of between 15 and 300 m. Turgo turbines are a variant on the Pelton wheel. It has a double cup in place of a single row of shallow cups of the Pelton, where water enters on one side and leaves on the other side. The Turgo turbine runner resembles a Pelton runner, but split in half. Therefore, for the same power output, the Turgo runner is one-half the size of the Pelton runner, implying that it has twice the specific speed of the Pelton. This specific speed of the turbine may be increased by increasing the number of jets which increases the runner, which increases the root of the number of jets, such that 4 more jets produce twice the specific speed of one jet on the same turbine [3, 38].

The Turgo turbines can handle larger volumes of water compared with a Pelton wheel of the same diameter, which makes them more suited for application in medium water heads [17, 38], while in terms of specific speed, the Turgo turbine runners speed lies between the specific speed of the Pelton and that of Francis turbines. Therefore, the Turgo turbine handles more water flow than the Pelton because the water exiting does not interfere with flow in adjacent buckets. The turbine can use single or multiple nozzles [38].

The Turgo turbine was designed in 1919 by Gilke with features that make it superior to the Francis and Pelton designs in some medium-head applications because of the following advantages;

i. The runner is cheaper than that of the Pelton wheel.

ii. It does not need an airtight housing, unlike the Francis turbine.

iii. With higher specific speed, the Turgo turbine has the capacity to handle a greater flow compared with the Pelton wheel of the same diameter, leading to lower generator and installation costs.

4.12 Cross-flow turbine/Bánki-Michell turbine/Ossberger turbine

The cross-flow turbine is an impulse-type turbine where water does not enter as a jet as in the Pelton turbine, but instead it enters as a flat sheet guided onto the blades and travels across the turbine and meets the blades again [17]. Anthony Michell, an Austrian engineer, designed the cross-flow turbine in the early 1900s. It was later improved by Donát Bánki, a Hungarian engineer, then further improved by Fritz Ossberger, a German engineer. Cross-flow turbine also called the Bánki-Michell or the Ossberger turbine provides a low-cost solution, especially in run-of-river hydro, with simplicity in construction being its salient feature [29, 39].

A cross-flow turbine, the Bánki-Michell turbine, or Ossberger was developed by the Australian Anthony Michell, the Hungarian Donát Bánki, and the German Fritz Ossberger. Michell patented the turbine design in 1903, and it was manufactured for many years by Weymouth. It was in 1933 that Ossberger's first patent was obtained. The patents related to this turbine are ("Free Jet Turbine" 1922, Imperial Patent No. 361593, and the "Cross Flow Turbine" 1933, Imperial Patent No. 615445) were manufactured as a standard product. The company founded by Ossberger, bearing his name, remains the leading manufacturer of this type of turbine today [29].

In this turbine, the water flows through the blades two times/twice. In the first pass, the water moves from outside of the blades to the inside. This is followed by a second pass flow where the water flows inside back out. The cross-flow turbine has fixed guide blades located at the entrance to the turbine for directing water to flow into a limited portion of the runner. The cross-flow turbine is designed to accommodate larger water flows and lower heads of water compared with the Pelton turbine [28]. The cross-flow turbine resembles a squirrel cage blower. It is drum-shaped with an elongated, rectangular section nozzle directed against curved vanes on a cylindrically shaped runner [39]. The turbine also runs clean as water leaves the runner, with grass, etc., that may reduce efficiency will not remain in the runner, hence no need for runner cleaning, a characteristic that compensates for the turbine's lower efficiency as it is more reliable than other types, which are often clogged more easily and consequently face power losses despite higher nominal efficiencies [3].

4.13 Deriaz pump-turbine

The Deriaz hydraulic turbomachine was developed by engineer Paul Deriaz as the first diagonal pump-turbine. In this turbine the flow is not full axial or full radial direction but a diagonal mixture of radial and axial flow. By having adjustable blades, the Deriaz turbine is to the Francis turbine what the Kaplan turbine is to the fixed-blade propeller. The turbine has reduced specific speed, leading to lower radial extension at the leading edge of the runner and loss of axial alignment to give a direction mix of centripetal and axial. To support increased hydraulic forces applied to the blades, the trunnions in the Deriaz turbine are larger and more resistant. The hydraulics in the Deriaz Pump-Turbine require more blades with an increase in the ratio of blade length to spacing, making the placement of the large pivot bearings very difficult. As a result of more energy difference between outlet and inlet, it becomes necessary to have longer blades to avoid blade overloading [40]. The length of movable blades is limited at the root by the hub's spherical geometry, which further limits the number of blades. In the Deriaz Pump-Turbine, the maximum number of blades is 10 which reduces maximum head and reduces maximum efficiency.

The runner blade servomotor for the Deriaz Pump-Turbine is often a rotary type located inside the turbine runner hub. The rotation of the servomotor piston through the piston rod is transmitted by the spider inside the runner hub. Connections to the spider and blade arm are equipped with slide blocks to convey the rotating motion of the spider to the runner blade stem.

The advantage of a Deriaz Pump-Turbine is the mechanical and hence a robust solution for pumped hydroelectric stations (PHES). Since the Deriaz Pump-Turbine has a wide operational range at synchronous speed without additional electrical equipment, it is less subject to failure in regulating power during charging and generating phases [2, 40].

4.14 PIT-type turbine-generators

PIT turbines are characterized by a vertical design, and their use saves a significant amount of space, making them ideal for small installations and hence are typically applied in low- to medium-head sites. They integrate quite well with existing dam structures without any need for extensive modifications. The pit turbine is a variation of the Bulb turbine, widely applied at a head range of 1.5–10 m and below a capacity of 15 MW. PIT-type turbine-generators are used for smaller, low-head applications, and they use a step-up gear for coupling it to a horizontal turbine shaft [26, 41]. PIT turbine-generators are ideal as an economical solution for low-head hydropower stations with lower ratings, typically up to 17 MVA, operating between 500 and 900 rpm, but with the ability to cope with high runaway speeds [36, 42]. The generator (with a speed-increasing gearbox) is contained within the upstream pit [43].

The pit generator turbine configuration has easy access to all the components, particularly the coupling of the turbine, speed increaser, and the generator, making inspection, maintenance, and repair easy. The pit turbine generator unit has higher efficiency than the S-type tubular turbine unit and is cheaper than the Bulb tubular unit [41, 43].

5. Sustainability of hydropower generation

Hydropower is generally regarded as a clean and reliable energy source, but it faces several challenges to its deployment and growth. Environmental impacts include interference with wildlife and natural habitats, displacement of people from their habitat, and alteration of the landscape. Technical challenges include the need to update infrastructure and the possibility of system failures. Large hydro schemes face financial problems in terms of financing [44, 45]. Concerns about the sustainability of hydropower have become significant and have to be understood and considered [46, 47].

Sustainable hydropower implies electricity generation from water in a manner that achieves a balance between environmental impacts, habitat preservation, and societal benefit. Hydropower remains the backbone of low-carbon power generation, about half of it globally, and about 55% higher than nuclear power contribution. Hydropower accounted for 17% of global electricity generation, making it the third-largest source of electricity after coal and natural gas, having grown by 70% over a period of 20 years [48, 49]. Hydropower supplies the majority of electricity demand in 28 countries, mainly in emerging and developing economies, having a total population of over 800 million. In developed countries, the share of hydropower in power generation has been on a decline due to aging power plants, with North America having an average hydropower plant age of 50 years old and 45 years old in Europe. These old plants need modernization to work effectively and play a role in sustainable power generation.

5.1 Environmental sustainability

Hydropower provides a valuable pathway for the energy transition because of its potential for carbon emissions reductions since its generation is associated with limited greenhouse gas emissions, as much as less than 1% of emissions from the most polluting coal power plants [50]. Hydropower is the leading source of global renewable power but still includes some hazardous emissions and adverse environmental impacts that should be mitigated [51]. The conventional hydropower system which uses storage by building dams to create reservoirs, has significant impacts compared to the run-of-river (or "hydrokinetic") systems, where water is diverted from rivers to spin hydraulic turbines [52]. Extreme weather conditions like droughts and heavy precipitation, which are expected to increase because of the climate change, will have profound effect on water resources and hence hydropower sustainability, which will also affect the economy, power production and use, human health, agriculture, transportation, national security, natural ecosystems, and recreation [51, 53].

Hydropower has some emissions, and hence it is not 100% emission-free since reservoirs created by dammed rivers do release some greenhouse gases like carbon dioxide and methane, mainly due to the decomposition of dead plants and other organic materials in the reservoir. Emissions can also be released by heating and cooling systems in hydropower plants, with the level of emissions varying with regional factors and site-specific factors [20].

Hydropower plants, particularly the run-of-the-river type, release far less greenhouse gas emissions when compared with the combustion of fossil fuels. Most emissions associated with hydropower occur during the construction, but with almost zero emissions during the operation of the plant itself, making run-of-the-river hydroelectric plants more beneficial to impoundment dams in terms of a smaller environmental footprint. Other environmental impacts include manipulation of river flows, which may lead to significant environmental impacts like negative effects on aquatic ecosystems affecting fish populations. This can be mitigated by the construction of fish ladders over small head differences in run-of-the-river systems that facilitate fish movement without harming them or interrupting their migration routes. Change in river basin or water composition may lead to higher species mortality, may cause an imbalance in biodiversity, and disrupt migration. Thermal pollution, if any, increases water turbidity at exit. It is, however, hard to establish whether or not the environmental damage is more than the benefits hence each project should be evaluated individually on a project-by-project basis [54].

Hydropower represents one of the oldest sources of energy for the generation of mechanical and electrical power and was the largest source of annual US renewable electricity generation up to 2019. Hydropower was used to turn paddle wheels on rivers to grind grain thousands of years ago. Hydropower was also used to power lumber and grain mills before steam power and electricity. Hydropower was first used to generate electricity at an industrial scale in the US in 1880, where it was used to light up 16 brush-arc lamps at the Wolverine Chair Factory located in Grand Rapids, Michigan. The first commercial. Hydroelectric power plant was used to sell electricity at the Fox River near Appleton, Wisconsin, on September 30, 1882 [55].

Hydropower turbines kill and injure fish that pass through the turbine, but well-designed turbines. Can reduce fish deaths to below 2%, compared with fish deaths of 5–10% for the best existing turbines. For fish like salmon and shad that swim upstream in rivers and streams for reproduction in spawning grounds in streams and rivers will have their movement interfered with by dam construction. Construction of *fish ladders* and elevators help the fish to move around or over dams to the spawning grounds upstream for reproduction [55, 56].

5.2 Social sustainability of hydropower

There are several social impacts of hydroelectric power plants besides the environmental impact. They include displacement of local populations through forced or willing land acquisitions for reservoir construction. They may significantly impair livelihoods and violate human rights [3, 20]. Hydropower serves as a source of both energy/electricity and water resource services and also helps to support other variable renewables, wind and solar, through storage and load balancing role [50, 57]. Dams are often developed with the assumption that locals will enjoy several benefits from them. However, hydropower development is associated with conflicts and controversies like the application of repression, criminalization, violence against activists, and even assassinations. Communities complain about the lack of consultation and attention to obvious negative impacts to the community and its environment. Common expectations and hence promises include cheaper energy bills, local job opportunities, better infrastructure, good schools and hospitals for better health care. There is a need for better benefit-sharing mechanisms, including compensations, with local communities, like sharing revenues from dams with municipalities [3, 58, 59].

Globally, millions of people are affected by dam construction, mainly due to permanent resettlement and disruptions of the people's livelihoods due to the construction and operation of dams. Compensation is necessary, but displaced people may get an undervalued price for their land or buildings with no or limited consideration of the social, cultural, and religious value of their land and the existing way that people use to make their livelihoods on the land or the river. Resettled people often lose their social networks and social wealth, which has significant economic, sociocultural, and health implications [6, 60, 61].

The limitations of current building practices can be overcome by incorporating regional to national policies that affect local issues in the design of dams and the information made available to stakeholders in a transparent manner. There is inadequate regional to multinational planning that considers dam impacts linked to the ecosystems. Assessments should be improved by incorporating the community concerns and to design new dams that can improve livelihoods by increasing crop productivity, maintain fisheries yields, increase food security, and improve access to water and energy from the project and avoid cases of developing hydropower schemes whose costs are greater than the benefit, with studies showing that many large dams across the world should not have been built [3, 61].

Regular and controlled flow of water, vegetation enrichment, increased agriculture including livestock production, and biodiversity. Dams have an important role in obstructing the passage of fallen trees and other objects that could potentially harm people and other water users [62, 63]. Often, special wildlife crossing ramps are developed at these power plants, which protects both the animals and the water resources.

5.3 Technical sustainability

Hydropower plants have one of the highest conversion efficiencies among all energy sources of about 90% efficiency in terms of water-to-wire efficiency. Although hydro has relatively high initial capital requirements, it has a long useful, life span at extremely low operation and maintenance costs. The global hydropower capacity is close to 4100 TWh a^{-1} accounting for close to 17% of global annual electricity generation, yet there is still significant potential for further development, with technical potential of close to 16,000 TWh, out of which close to 9600 TWh is classified as economically viable. About 50% of technical potential in Europe has already been developed, while in Asia just 24% is developed and just 6% is already developed in Africa. Additionally, huge potential can be found in existing developed infrastructure lacking generating units, which can be utilized by adding new hydropower facilities. Only about 25% of close to 45,000 large dams globally are used for hydropower, while close to 75% are used exclusively for other purposes like irrigation, flood control, navigation, and urban water supply schemes [50]. Therefore, redevelopment to generate power can add huge quantities of green power to global grids.

Small dams are emerging as a more benign technology than large dams, although so far with limited evidence with the US alone. Building about 2 million low-head or small dams. Being abundant implies that small dams can substantially impact flowing aquatic ecosystems since small hydro dams share characteristics with large dams, other than their size, with India and China laddering in small hydro dams. They have significant benefits, particularly in remote and isolated mountainous terrain. Even small hydropower should be subjected to both environmental impact assessments and environmental impact reports, generally for those above 10 MW [64, 65].

The capacity factor for hydropower plants varies over time and across installations based on hydrological patterns and plant design and purposes of the power plant, with most conventional projects operating between 25 and 80% range. The capacity factor of a hydropower plant is determined by many factors like energy market factors and conditions, hydrological conditions, and a number of plant-level variables like run-of-river, reservoir, or PSH, which is designed to operate during the peak or baseload or frequency response, power outage schedules, resource quality, minimum ecological flow requirements, and multiple-usage constraints [36, 66].

In terms of flexibility, hydropower plants have one of the highest technical capabilities for providing operational flexibility. In most applications, hydropower offers balancing and ancillary services like reserve regulation and quick frequency response in primary and secondary frequency regulation [36]. Technological innovations and digitalization in the future may make power plants operate more efficiently by precise and timely flow regulation, while power plant modernization and upgrading and the construction of entirely new plants at the numerous viable sites globally will increase the role and contribution of hydropower in the energy transition [62].

5.4 Institutional or political sustainability

Hydropower generation needs consideration of governance implications of the dam construction and associated energy distribution and use. Policy makers should not just view energy as the entry point for hydroelectric schemes and use water as a means to produce without considering the effects on livelihoods. Policies should be conceived with a nexus approach, particularly the food-water-energy nexus, whose implications go beyond one country, e.g., the construction of binational hydroelectric dams on the Bolivia/Brazil border, where flooding in one country leads to flooding in the neighboring country at the downstream of a river [61].

Accidents do occur in hydropower plants mainly due to equipment failure, poor planning, or unpredictable natural events. New dam developers may also fail to take into consideration critical geological factors. Although accidents are rare in hydropower plants, they may have fatal consequences when they happen; hence the need for planners and designers to seriously consider the geological and environmental issues during design, planning, and construction, and the state or regulators should ensure that hydropower infrastructure and its water resources should be subjected to Sustainability Assessment Protocol. The Hydropower Sustainability Assessment which is a tool for assessing projects across a range of social, environmental, technical, and economic dimension [63].

It is important for the state and regulators to be actively involved in hydropower development beyond licensing to include overall management of water resources and all its activities to protect the water resources from any harm associated with hydropower development and to ensure sustainable management of water resources and ensure sustainable power generation and maximum protection of water resources and its infrastructure [63].

5.5 Economic sustainability of hydropower

The construction of hydropower plants has a relatively high initial cost, but in the long to medium term, hydropower generation is economically advantageous. Hydroelectric power systems are simple to operate as a storage system; the average yield of the process is about 70%, which is difficult to achieve with traditional sources. Hydropower plants do not consume water, as the resources used is returned in their entirety to the natural course of water [62].

Hydropower is economically competitive, having the levelized cost of energy (LCOE) typically between 2 and 10 US c$\cdot$(kWh)$^{-1}$, which is comparable with that of athermal power plants. Hydropower also has an attractive payback ratio of 200–300, which is the highest of all types of renewable energies [50, 67]. There is a complex relationship between hydropower and economic growth and financial development involving many factors. Hydroelectric power generation may positively affect economic growth and contribute to financial development [68]. Being the least expensive large-scale energy storage technology today, having lower operational costs makes it the least costly source of energy storage for many energy markets and countries. Developers, however, face a trade-off, choosing either to maximize revenue through electricity generation or to use storage capabilities to provide flexibility and ancillary services, all of which are beneficial to the grid and hence the society [36].

6. Summary

Hydropower is widely used as a source of clean energy, but it also emits hazardous emissions, having a negative effect on human health, ecosystems, and resources. Hydropower benefits include their relatively low acquisition cost against durability, enabling energy independence, high flexibility, and adaptability for spatial conditions and user energy demands. Micro-hydro systems could capitalize on head range as low as 100 cm, and while the efficiency of hydroelectric systems varies between 65 and 75% in micro and small applications and as high as more than 96% in macro designs. Hydropower is attractive mostly because of its flexibility in applications, adaptability to combine with other renewable energy systems, and the ability to produce clean, stable, and reliable energy.

Globally we have close to 16,000 TWh as the total technical potential for further development, of which about 9600 TWh is classified as economically viable. The annual growth in capacity remains minimal, with different regions having varying growth potential, with Europe having more than 50% of technical potential already developed, while Asia is at 24% and Africa just 6% of the technical growth potential has been achieved. Additionally, significant potential still exists in developed hydro infrastructure lacking generating units, which can be exploited by adding generating infrastructure. Over 75% of the existing 45,000 large dams in the world are not used for hydropower but are used exclusively for other purposes like irrigation, flood control, navigation, and urban water supply schemes.

Hydropower is abundantly available and is one of the most efficient energy resources. Hydropower transforms the potential energy of stored water at a height to spin a turbine by the action of water falling on the blades of the turbine. The various classes of hydroelectric power plants include dammed reservoir, run-of-river, and pumped storage. Hydropower plants can also be classified into small, mini, micro, and large hydropower plants on the basis of size [9, 69].

Turbine type	Features and operation	Applications
Reaction turbine	Water at high pressure enters the turbine tangentially to the runner via the guide vanes, which are fixed, then directed to adjustable guide vanes at a high tangential velocity component.	Low to medium water heads and higher water discharge.
Impulse turbine	The impulse turbines efficiently convert the kinetic energy of moving in the form of high-velocity water jets into mechanical energy in the form of turbine shaft work using bucket-shaped blades.	High water head applications at low flow rates or discharge.
Tubular turbine	Tubular or tube turbines are horizontal or slant-mounted turbines equipped with propeller runners, with the generators located outside of the water passageway. The turbines have fixed or variable pitch runners and with or without wicket gate assemblies.	Tubular turbines are suitable for very low hydraulic head. These turns feature a straight flow-through design that enables the water to flow linearly through the turbine. They are often applied in tidal and wave power plants.
Kaplan turbine	The Kaplan turbine is an axial flow suitable for locations with low head and high water flow rates characterized by adjustable blades, which helps in optimizing under varying water conditions to maintain high efficiency.	The Kaplan turbine is ideal for medium-head schemes and is particularly useful in river and run-of-river hydroelectric schemes.
Propeller turbine	The propeller turbine has a runner with three to six fixed or adjustable blades and is operated as water constantly contacts all the blades.	The propeller turbine is ideal for medium-head hydroelectric schemes.
Bulb	The Bulb turbines are horizontal-axis turbines with propeller runners directly connected to the generator, which is enclosed in a watertight enclosure (bulb) located in the turbine water passageway. The turbine has got either fixed or variable pitch blades with a wicket gate mechanism or none. The Bulb turbine has a relatively higher efficiency of about 2% higher than the vertical unit and 1% more than the tube unit because of the straight water passageway.	They have lower space requirements because of the compact design but require longer time during maintenance because of accessibility challenge; however, they may be greater than for either the vertical or the tube-type turbines.
Straflo turbine	Straflo turbines, also called the straight flow turbines, are axial turbine-generators located outside the water delivery tube connected to the outer ring on the edge of the rotor blades. A sealing is installed to limit water leakage from the rotor blades to the generator.	The Straflo has a generator around the periphery of the turbine's runner, which enables direct connection between the generator and the turbine, which enhances the overall efficiency and compactness. This turbine is ideal for low-head hydroelectric plants.
Francis	The Francis turbine has a runner with generally nine or more fixed blades, more blades in which water is introduced above the runner and then all around it before it falls through, forcing the blades to rotate. The Francis turbine is a radial-flow turbine in which water flows inwards toward the center. The turbines can achieve as high as 95% efficiency under optimum conditions in terms of speed and direction of the incoming water relative to the runner blades.	The Francis is a preferred choice for medium-head installations covering a wide range of flow and head conditions. The turbine is a radial inflow design with water entering the runner radially while exiting axially. The Francis turbine is the most used turbine globally due to its flexibility.

Turbine type	Features and operation	Applications
Kinetic turbines	Kinetic energy turbines, also called free-flow turbines, generate power by converting kinetic energy of moving water instead in rivers, man-made channels, tidal waters, or ocean water currents. They do not need a diversion of water through pipes, riverbeds, or man-made channels, even though they can still be used. These turbines are portable, which reduces the need for costly infrastructure. They are designed to be installed underwater, in floating, fixed, anchored, or towed configurations, and at any place where water flows at a minimum speed of about 0.25 m/s.	These turbines are installed along the canals, having suitable conditions for flow volumes, velocities, and flow reliability. These turbines allow new applications and new potential in previously unexplored, long, flat river or canal sections where conventional hydro turbines do not apply with very minimal civil works. Hydrokinetic turbines can supply base load for the microgrid power systems.
Pelton	The Pelton wheel is the oldest impulse turbine, applied in high-head, low flow water conditions. The Pelton turbine is the most popular impulse turbine suitable for high water heads of 200–4000 m but low water flow rates of between 0.5 and 20 m^3/s and net power of up to 200 MW, at an efficiency of about 90%.	The turbine consists of spoon-shaped buckets installed on the wheel's edge, meant to capture water jets exiting nozzles, which turns the wheel in a highly efficient process.
Turgo turbines	The Turgo turbine is an impulse turbine. The Turgo turbine is a product of the modification of the Pelton wheel. This is an impulse-type turbine used for medium-head applications and achieves efficiencies of about 87–90% operating at net heads of between 15 m and 300 m. Turgo turbines are a variant on the Pelton wheel.	In which a water jet is directed at an angle on the runner, which facilitates handing of more flow at a lower head as compared to the Pelton wheel. The Turgo turbine is more compact and often applied in medium-head installations.
Cross-flow turbine	This is an impulse-type turbine where water does not enter as a jet as in the Pelton turbine but as a flat sheet guided onto the blades and travels across the turbine and meets the blades again. Anthony Michell, an Austrian engineer, designed the cross-flow turbine in the early 1900s. It was later improved by Donát Bánki, a Hungarian engineer, then further improved by Fritz Ossberger, a German engineer. This turbine provides a low-cost solution, especially in run-of-river hydro, with simplicity in construction being its salient feature.	The cross-flow/Bánki-Michell turbine has a simple design in which water flows through a rectangular section runner twice. Leading to higher efficiency across a range of head and flow conditions. The simplicity in design makes the cross-flow turbine ideal for small-scale applications like in community-based hydro projects.
Deriaz Pump-Turbine	This turbine is a diagonal pump-turbine. In which the water flows is not full axial or full radial direction but a diagonal mixture of radial and axial flow. By having adjustable blades, the Deriaz turbine, is to the Francis turbine what the Kaplan turbine is to the fixed-blade propeller. The Deriaz hydraulic turbomachine was developed by engineer Paul Deriaz as the first diagonal pump-turbine. In this turbine the flow is not full axial or full radial direction but a diagonal mixture of radial and axial flow.	By having adjustable blades, the Deriaz turbines is to the Francis turbine what the Kaplan turbine is to the fixed-blade propeller. The turbine has reduced specific speed, leading to lower radial extension at the leading edge of the runner and loss of axial alignment to give a direction mix of centripetal and axial.

Turbine type	Features and operation	Applications
PIT turbine	For the small hydropower station with a low water head, the shaft-extension tubular type turbine and pit turbine show equivalent technical and economic advantages to the Bulb tubular turbine.	The PIT turbines can operate with the lowest water head of conventional pit turbine units but generally above 4 meters with design discharge of larger than 20 m^3/s and an efficiency of close to 87%, but the highest efficiency is 90%.

Table 2.
Hydroelectric turbines, features, and applications.

	Sustainability dimension	Impacts of hydropower generation
1	Environmental	Interruption of water life, including fish, greenhouse gas emissions from dams. Hydropower has some emissions, and hence it is not 100% emission-free since reservoirs created by dammed rivers do release some greenhouse gases like carbon dioxide and methane, mainly due to the decomposition of dead plants and other organic materials in the reservoir. Emissions can also be released by heating and cooling systems in hydropower plants, with the level of emissions varying with regional factors and site-specific factors.
2	Social	Forced displacement and interruption of the social fabric. Can also lead to human rights abuses during displacement. Globally, millions of people are affected by dam construction, mainly due to permanent resettlement and disruptions of the people's livelihoods due to the construction and operation of dams.
3	Technical	In terms of flexibility, hydropower plants have one of the highest technical capabilities for providing operational flexibility. In most applications, hydropower offers balancing and ancillary services like reserve regulation and quick frequency response in primary and secondary frequency regulation.
4	Economic sustainability	Can enhance the local economy through irrigation, fish farming, and tourism. The construction of hydropower plants has a relatively high initial cost, but in the long to medium term, hydropower generation is economically advantageous. Hydroelectric power systems are simple to operate as a storage system; the average yield of the process is about 70%, which is difficult to achieve with traditional sources. Hydropower is economically competitive, having the levelized cost of energy (LCOE) typically between of 2 and 10 US c·$(kWh)^{-1}$, which is comparable with that of thermal power plants. Hydropower also has an attractive payback ratio of 200:300, which is the highest of all types of renewable energy.
5	Institutional/political sustainability	Accidents do occur in hydropower plants mainly due to equipment failure, poor planning, or unpredictable natural events. New dam developers may also fail to take into consideration critical geological factors. Although accidents are rare in hydropower plants, they may have fatal consequences when they happen; hence the need for planners and designers to seriously consider the geological and environmental issues during design, planning, and construction, and the state or regulators should ensure that hydropower infrastructure and its water resources should be subjected to Sustainability Assessment Protocol.

Table 3.
Sustainability of hydropower generation.

There is a need to displace fossil fuels rapidly with cleaner fuels and energy sources in order to stabilize the climate. Hydropower currently accounts for about two-thirds of renewable energy sources globally and is still expected to reach 45% by 2040. Analyses of climate impacts of hydropower plants have been emphasizing the aggregated 100-year impacts from a one-year pulse of emissions. Through analysis of climate impacts of sustained hydropower emissions over time, it is established that there are enormous differences in climate impact amounts. There may be limited or no benefits if climate impact minimization is not prioritized in the power plant design and construction of new hydropower facilities, which may lead to limited or even no climate benefits [20, 66].

Hydropower generation creates several environmental impacts, economic impacts, social and legislative as well as institutional challenges, and technical challenges and opportunities, which have been analyzed in this book chapter. The impacts largely depend on specific characteristics of installation, benefits transfer techniques, and existing policy, legal, and legislative frameworks. This chapter demonstrates a case study approach in identifying and mitigating potential impacts in developing and exploiting hydropower [70].

6.1 Summary of turbine classifications

Hydraulic turbines convert the energy of flowing water, the pressure, or both into mechanical work that spins a generator for power generation. By varying the turbine design and application, the specific hydraulic turbine is suitable for specific purposes based on factors like the water flow, the head, and power demand. **Table 2** summarizes the various turbines, characteristics, and applications.

From **Table 2**, it is noted that there are various hydraulic turbines that can be selected and used over a wide variety of hydropower plants that vary in specific site conditions, operational efficiency and characteristics, size, and output capabilities. This shows that hydropower is technically mature and sustainable.

6.2 Summary of sustainability assessment

From **Table 3**, it is noted that hydropower has sustainability challenges that should be addressed and opportunities to be exploited for it to play a critical role in the energy transition.

7. Conclusion

Hydropower is the main source of renewable energy, providing other benefits like water and energy storage and flexibility. On the negative side, hydropower may cause significant impacts to the environment. Therefore, for hydropower to be sustainable, it should realize a balance between power generation, environmental impacts, and benefits to society, aiding the achievement of the emissions and climate targets. Hydropower is said to be sustainable when electricity production from water realizes a balance between environmental impacts, habitat preservation, and societal benefit.

Hydropower plants can be classified into three types of facilities, namely impoundment, *diversion, and pumped storage*. Dams are very useful components of hydropower by forming storage that makes hydropower dispatchable plants. However, for several reasons, not all hydroelectric power schemes use dams. Hydropower is the largest source of clean, renewable electricity, but just about 30% of the global potential of about 20 billion MWh per year has been developed so far, accounting for close to 17% of global electricity generation with an annual generation capacity of about 4100 TWh a^{-1} globally. However, the share or contribution to the energy mix for hydropower has been reducing over time, mainly because of rapid growth in the share of wind and solar energy. Hydropower is a proven, mature technology that is environmentally friendly, price competitive, flexible, and has a high conversion efficiency of more than 90% in terms of water-to-wire and is socially acceptable in many cases, even though it is associated with negative impacts like the displacement of people, negative effects on aquatic life, and emissions from large dams.

The use of hydropower provides a significant potential for reductions in global greenhouse gas emissions, which are typically less than 1% of emissions typical conventional coal power plants. Hydropower is also cost and price having a levelized cost of energy (LCOE) typically in the range of 2–10 US c $(kWh)^{-1}$, which is comparable with the energy payback ratio of 200:300, which is the highest of all types of renewable energies. Other than power generation, hydropower facilities supply water services and support with storage and load balancing services.

Hydropower is a clean renewable energy resource that can minimize greenhouse gas emissions and global warming effects. More use of hydropower will limit or avoid the use of fossil fuels in power generation and thus leave a cleaner world. Hydropower is a no exhaustible natural resource maintained by natural forces in the water cycle, making it a long-lasting source of electric power and water for multiple applications, which further increases its sustainability. It requires a relatively high initial investment but has a long life span with very low operation and maintenance costs.

Author details

Moses Jeremiah Barasa Kabeyi* and Oludolapo Akanni Olanrewaju
Industrial Engineering Department, Durban University of Technology, Durban, South Africa

*Address all correspondence to: moseskabeyi@yahoo.com and mosesk1@dut.ac.za

References

[1] Unya IF, Sodiki JI, Nkoi B. Performance evaluation of a hydro-power plant. International Journal for Research in Applied Science and Engineering Technology (IJRASET). 2022;**10**(XII):2376-2385. DOI: 10.22214/ijraset.2022.48162

[2] Kabeyi MJB, Olanrewaju OA. Hydropower in the sustainable energy mix. In: Second Australian International Conference on Industrial Engineering and Operations Management, Melbourne, Australia, November 14-16, 2023. 2023. p. 247. Available from: https://ieomsociety.org/proceedings/2023australia/247.pdf

[3] Kabeyi MJB. Sustainable Energy Transition and Optimization of Grid Electricity Generation and Supply [Doctor of Engineering Thesis, Industrial Engineering]. Durban, South Africa: Durban University of Technology; 2024. Available from: https://hdl.handle.net/10321/5485

[4] Kabeyi MJB, Olanrewaju OA. Sustainable energy transition for renewable and low carbon grid electricity generation and supply, (in English). Frontiers in Energy Research, Review. 2022;**9**(743114):1-45. DOI: 10.3389/fenrg.2021.743114

[5] Hunt JD et al. Hydropower and seasonal pumped hydropower storage in the Indus basin: Pros and cons. Journal of Energy Storage. 2021;**41**:102916. DOI: 10.1016/j.est.2021.102916

[6] Adu D, Jianguo D, Darko RO, Boamah KB, Boateng EA. Investigating the state of renewable energy and concept of pump as turbine for energy generation development. Energy Reports. 2020;**6**:60-66. DOI: 10.1016/j.egyr.2020.08.025

[7] Mohamed T. Chapter ten - Hydropower. In: El Bassam N, editor. Distributed Renewable Energies for off-Grid Communities. 2nd ed. Boston: Elsevier; 2021. pp. 213-230

[8] Wagner H-J, Mathur J. Components of hydropower plants. In: Wagner H-J, Mathur J, editors. Introduction to Hydro Energy Systems: Basics, Technology and Operation. Berlin, Heidelberg: Springer Berlin Heidelberg; 2011. pp. 49-70

[9] Kumar K, Saini RP. A review on operation and maintenance of hydropower plants. Sustainable Energy Technologies and Assessments. 2022;**49**:101704. DOI: 10.1016/j.seta.2021.101704

[10] Kabeyi MJB, Olanrewaju AO. Hydropower In the sustainable energy mix. In: 6th European Conference on Industrial Engineering and Operations Management, Lisbon, Portugal, July 18-20, 2023. Amsterdam, The Netherlands; 2023. p. 247

[11] El Bassam N, Maegaard P, Schlichting ML. Chapter ten - Hydropower. In: El Bassam N, Maegaard P, Schlichting ML, editors. Distributed Renewable Energies for off-Grid Communities. Amsterdam, The Netherlands: Elsevier; 2013. pp. 167-174

[12] Energy Efficiency and Renewable Energy. U.S. Hydropower Market Report. Washinton DC: US Energy Department; 2021. Available from: https://www.energy.gov/sites/prod/files/2021/01/f82/us-hydropower-market-report-full-2021.pdf

[13] Gallagher J, Styles D, McNabola A, Williams AP. Life cycle environmental balance and greenhouse gas mitigation

potential of micro-hydropower energy recovery in the water industry. Journal of Cleaner Production. 2015;**99**:152-159. DOI: 10.1016/j.jclepro.2015.03.011

[14] Arnaiz M, Cochrane TA, Calizaya A, Shrestha M. A framework for evaluating the current level of success of micro-hydropower schemes in remote communities of developing countries. Energy for Sustainable Development. 2018;**44**:55-63. DOI: 10.1016/j.esd.2018.03.002

[15] Killingtveit Å. Chapter 21 - Hydroelectric power. In: Letcher TM, editor. Future Energy. 2nd ed. Boston: Elsevier; 2014. pp. 453-470

[16] Tkáč Š. Hydro power plants, an overview of the current types and technology. Selected Scientific Papers - Journal of Civil Engineering. 2018;**13**(s1):115-126. DOI: 10.1515/sspjce-2018-0011

[17] Enel Green Power. Hydropower. Rome, Italy: Enel Greenpower; 2024. Available from: https://www.enelgreenpower.com/learning-hub/renewable-energies/hydroelectric-energy/hydroelectric-plants [Accessed: September 18, 2024]

[18] U.S. Department of Energy. Hydropower Basics. Washington, DC: The U.S. Department of Energy; 2024 https://www.energy.gov/eere/water/hydropower-basics [Accessed: September 15, 2014]

[19] Energy Efficiency and Renewable Energy. Types of Hydropower Plants. Washington, DC: Energy Efficiency Office; 2023. Available from: https://www.energy.gov/eere/water/types-hydropower-plants [Accessed: February 27, 2023]

[20] Kuriqi A, Pinheiro AN, Sordo-Ward A, Bejarano MD, Garrote L. Ecological impacts of run-of-river hydropower plants—Current status and future prospects on the brink of energy transition. Renewable and Sustainable Energy Reviews. 2021;**142**:110833. DOI: 10.1016/j.rser.2021.110833

[21] Sharma H, Singh J. Run off river plant: Status and prospects. International Journal of Innovative Technology and Exploring Engineering (IJITEE). 2013;**3**(2):210-213. Available from: https://www.ijitee.org/wp-content/uploads/papers/v3i2/B0971073213.pdf

[22] Hegde RK. Power plant engineering. Uttar Pradesh, India: Pearson India Education Services Pvt. Ltd.; 2015

[23] Energy Efficiency and Renewable Energy. Pumped Storage Hydropower. Washington, DC: US Department of Energy; 2023. Available from: https://www.energy.gov/eere/water/pumped-storage-hydropower [Accessed: February 27, 2023]

[24] Rehman S, Al-Hadhrami LM, Alam MM. Pumped hydro energy storage system: A technological review. Renewable and Sustainable Energy Reviews. 2015;**44**:586-598. DOI: 10.1016/j.rser.2014.12.040

[25] Sadi M, Rahbari HR, Singh AK, Arabkoohsar A. 10 - Pumped hydropower energy storage. In: Arabkoohsar A, editor. Future Grid-Scale Energy Storage Solutions. Cambridge, Massachusetts, USA: Academic Press; 2023. pp. 409-433

[26] Yang C, Zheng Y, Li L. Optimization design and performance analysis of a pit turbine with ultralow head. Advances in Mechanical Engineering. 2014;**6**:401093. DOI: 10.1155/2014/401093

[27] Dincer I, Ezzat MF. 3.4 renewable energy production. In: Dincer I, editor.

Comprehensive Energy Systems. Oxford: Elsevier; 2018. pp. 126-207

[28] Energy Efficiency and Renewable Energy. Types of Hydropower Turbines. Washington, DC: US Energy Department; Available from: https://www.energy.gov/eere/water/types-hydropower-turbines; [Accessed: July 27, 2023]

[29] Jawahar CP, Michael PA. A review on turbines for micro hydro power plant. Renewable and Sustainable Energy Reviews. 2017;**72**:882-887. DOI: 10.1016/j.rser.2017.01.133

[30] Suneco Hydro. Tubular Turbine. Suneco Hydro. Available from: https://www.micro-hydro-power.com/tubular-turbine-bulb-type.htm; [Accessed: November 05, 2024]

[31] Vagnoni E. The reaction turbines used in PHES units and related problems. In: Cabeza LF, editor. Encyclopedia of Energy Storage. Oxford: Elsevier; 2022. pp. 123-135

[32] Ebrahimi M. 12 - Hydro power plants. In: Ebrahimi M, editor. Power Generation Technologies. Cambridge, Massachusetts, USA: Academic Press; 2023. pp. 495-531

[33] Hanif MA, Nadeem F, Tariq R, Rashid U. Chapter 7 - Power generation by ocean energy. In: Hanif MA, Nadeem F, Tariq R, Rashid U, editors. Renewable and Alternative Energy Resources. Cambridge, Massachusetts, USA: Academic Press; 2022. pp. 359-430

[34] Singh AR, Prasad V. Design and performance analysis of Straflo turbine using CFD. IOSR Journal of Mechanical and Civil Engineering (IOSR-JMCE). 2016;**13**:54-59. DOI: 10.9790/1684-1304035459

[35] Singh A, Prasad V. Design and flow simulation of Straflo turbine for micro hydro Power Station. In: 6th International and 43rd National Conference on Fluid Mechanics and Fluid Power, December 15-17, 2016. 2016. p. 295 Available from: https://www.scribd.com/document/507279762/fmfp

[36] IEA. Hydropower Special Market Report. France: International Energy Agency Paris; 2021. Available from: https://www.iea.org/reports/hydropower-special-market-report

[37] Sher F et al. Emerging technologies for biogas production: A critical review on recent progress, challenges and future perspectives. Process Safety and Environmental Protection. 2024;**188**:834-859. DOI: 10.1016/j.psep.2024.05.138

[38] Benzon DS, Aggidis GA, Anagnostopoulos JS. Development of the Turgo impulse turbine: Past and present. Applied Energy. 2016;**166**:1-18. DOI: 10.1016/j.apenergy.2015.12.091

[39] Anand RS, Jawahar CP, Bellos E, Malmquist A. A comprehensive review on crossflow turbine for hydropower applications. Ocean Engineering. 2021;**240**:110015. DOI: 10.1016/j.oceaneng.2021.110015

[40] Morabito A, de Oliveira e Silva G, Hendrick P. Deriaz pump-turbine for pumped hydro energy storage and micro applications. Journal of Energy Storage. 2019;**24**:100788. DOI: 10.1016/j.est.2019.100788

[41] L. Chongqing Hydropower Equipment Co. Pit Turbine. Vol. 2024. Chongqing City, China: Chongqing Hydropower Equipment Co., Ltd.; 2011. Available from: http://www.cchpe.net/Productinfo.aspx?get=2

[42] Kabeyi MJB, Olanrewaju OA. Environmental impacts of power

plants and energy conversion systems. In: Chen L, editor. Advances in Clean Energy Systems and Technologies. Cham: Springer Nature Switzerland; 2024. pp. 445-469

[43] Voith. Bulb/Pit/S-Turbines and Generators. Heidenheim, Germany: Voith Hydro Holding GmbH & Co. KG; 2024. Available from: https://voith.com/corp-en/Voith_t3366e_Bulb_Pit_S-turbines_screen.pdf

[44] Gracey EO, Verones F. Impacts from hydropower production on biodiversity in an LCA framework—Review and recommendations. The International Journal of Life Cycle Assessment. 2016;**21**(3):412-428. DOI: 10.1007/s11367-016-1039-3

[45] de Faria FAM, Davis A, Severnini E, Jaramillo P. The local socio-economic impacts of large hydropower plant development in a developing country. Energy Economics. 2017;**67**:533-544. DOI: 10.1016/j.eneco.2017.08.025

[46] Nautiyal H, Goel V. Sustainability assessment of hydropower projects. Journal of Cleaner Production. 2020;**265**:121661. DOI: 10.1016/j.jclepro.2020.121661

[47] Kabeyi MJB, Olanrewaju OA. Types of grid scale energy storage batteries. In: Chen L, editor. Advances in Clean Energy Systems and Technologies. Cham: Springer Nature Switzerland; 2024. pp. 181-203

[48] IEA. Hydropower Special Market Report Analysis and Forecast to 2030. Paris: International Energy Agency; 2021. Available from: https://iea.blob.core.windows.net/assets/4d2d4365-08c6-4171-9ea2-8549fabd1c8d/HydropowerSpecialMarketReport_corr.pdf

[49] Tefera WM, Kasiviswanathan KS. A global-scale hydropower potential assessment and feasibility evaluations. Water Resources and Economics. 2022;**38**:100198. DOI: 10.1016/j.wre.2022.100198

[50] Killingtveit Å. 15 - Hydroelectric power. In: Letcher TM, editor. Future Energy. 3rd ed. Amsterdam, The Netherlands: Elsevier; 2020. pp. 315-330

[51] Mahmud MAP, Huda N, Farjana SH, Lang C. A strategic impact assessment of hydropower plants in alpine and non-alpine areas of Europe. Applied Energy. 2019;**250**:198-214. DOI: 10.1016/j.apenergy.2019.05.007

[52] National Energy Technology Laboratory. Life Cycle Analysis (LCA) of Energy Technology and Pathways Life Cycle Analysis (LCA) of Energy Technology and Pathways. Pittsburgh, Pennsylvania: US Department of Energy; 2023. Available from: https://netl.doe.gov/LCA [Accessed: May 2, 2023]

[53] NOAA. Water Cycle. National Oceanic and Atmospheric Adminisration. Washington, DC. Available from: https://www.noaa.gov/education/resource-collections/freshwater/water-cycle; [Accessed: September 16, 2024]

[54] Song C, Wang G, Sun X, Hu Z. River runoff components change variably and respond differently to climate change in the Eurasian Arctic and Qinghai-Tibet plateau permafrost regions. Journal of Hydrology. 2021;**601**:126653. DOI: 10.1016/j.jhydrol.2021.126653

[55] US Energy Aministration Information. Hydropower Explained. Washington, D.C: US Energy Aministration Information; 2023. Available from: https://www.eia.gov/energyexplained/hydropower/ [Accessed: October 24, 2023]

[56] Kabeyi MJB, Olanweraju OA. Sustainability assessment for non-combustible renewable power generation. In: 12th Annual Istanbul International Conference on Industrial Engineering and Operations Management Istanbul, Turkey, March 7-10, 2022. 2022, Paper 429. Available from: https://ieomsociety.org/proceedings/2022istanbul/429.pdf

[57] Kabeyi MJB, Olanrewaju AO. Application of geothermal wellhead generators in sustainable power generation. Geothermal Resources Council Transactions. 2022;**46**(2022):1692-1718. Art no. 1034703. Available from:https://www.geothermal-library.org/index.php?mode=pubs&action=view&record=1034703

[58] Kabeyi MJB, Oludolapo O. Performance analysis and development of an export cogeneration plant for a 3000 TCD sugar cane factory. In: 12th Annual International Conference on Industrial Engineering and Operations Management, Istanbul, Turkey, March 7-10, 2022. 2022. p. 409 Available from: https://index.ieomsociety.org/index.cfm/article/view/ID/244

[59] Kabeyi MJB, Olanrewaju AO. Micro grids: Design, operation and applications. In: 4th Asia Pacific International Conference on Industrial Engineering and Operations Management Vietnam. 2023. Available from: https://ieomsociety.org/proceedings/2023vietnam/269.pdf

[60] IRENA. Hydropower technology brief. In: The International Renewable Energy Agency (IRENA). 2015. Available from: https://www.irena.org/-/media/Files/IRENA/Agency/Publication/2015/IRENA-ETSAP_Tech_Brief_E06_Hydropower.pdf

[61] Moran EF, Lopez MC, Moore N, Müller N, Hyndman DW. Sustainable hydropower in the 21st century. Proceedings of the National Academy of Sciences. 2018;**115**(47):11891-11898. DOI: 10.1073/pnas.1809426115

[62] Enel Green Power and Thermal Generation (EGP&TGX). Hydroelectric, the world's oldest renewable energy source, is a key player in today's energy transition. In: Enel Green Power and Thermal Generation (EGP&TGXPower). 09 November 2022. Available from: https://www.enelgreenpower.com/stories/articles/2022/11/hydroelectric-role-energy-transition. [Accessed: November 26, 2022]

[63] Draft Policy on Sustainable Hydropower Generation. Department of Water and Sanitation, Government of South Africa; 2016. Available from: https://static.pmg.org.za/160718draftsustainablehydropowergeneration.pdf

[64] Kabeyi MJB, Olanrewaju AO. Solar energy as a sustainable energy for power generation. In: 2nd Australian International Conference on Industrial Engineering and Operations Management. 2024. Available from: https://ieomsociety.org/proceedings/2023australia/244.pdf

[65] Kabeyi MJB, Olanrewaju AO. Wind energy in sustainable power generation and supply. In: 6th European Conference on Industrial Engineering and Operations Management, Lisbon, Portugal, July 18-20, 2023. 2023. p. 246

[66] Kabeyi MJB, Olanrewaju OA. Optimum energy potential of an operating digester for a slaughterhouse. AIP Conference Proceedings. 2023;**3018**(1):1-16. DOI: 10.1063/5.0171602

[67] Kabeyi MJB. Project and program evaluation consultancy with terms of reference, challenges, opportunities, and

recommendations. International Journal of Project Management and Productivity Assessment (IJPMPA). 2020;**8**(2):47-68. DOI: 10.4018/ijpmpa.2020070103

[68] Ullah A, Altay Topcu B, Dogan M, Imran M. Exploring the nexus among hydroelectric power generation, financial development, and economic growth: Evidence from the largest 10 hydroelectric power-generating countries. Energy Strategy Reviews. 2024;**52**:101339. DOI: 10.1016/j.esr.2024.101339

[69] He M, Han S, Gao Y, Zhao Z, Chen D, Deng M. Exploring the impact of three representative pumped storage retrofits on the economic-technical-energy efficiency of cascaded hydropower-VRE hybrid systems. Energy Conversion and Management. 2024;**322**:119107. DOI: 10.1016/j.enconman.2024.119107

[70] Botelho A, Ferreira P, Lima F, Pinto LMC, Sousa S. Assessment of the environmental impacts associated with hydropower. Renewable and Sustainable Energy Reviews. 2017;**70**:896-904. DOI: 10.1016/j.rser.2016.11.271

Chapter 5

Multi Use Water System Approach for Small Hydro Power Plants (SHPP) alongside Irrigation and Drinking Water Supply (DWS) Networks

Ahmed Moulaye Abdou and Adam Ahmed Abdou

Abstract

The multi-use water system (MUWS) approach means that different uses are considered at the planning stage, and the infrastructure is designed to support them. The implementation of hydro-agricultural irrigation schemes is an opportunity to develop multi-use water points. Alongside the development of irrigation, other productive activities can be developed (drinking water supply, livestock breeding, fish farming, hydroelectricity, etc.). Likewise, Multi-Village Drinking Water Supply (MV-DWS) projects offer a good opportunity for the development of irrigation, livestock, fish farming, and hydroelectricity. This chapter aims at the conceptualization of projects for the integrated development of irrigation, drinking water supply, and hydroelectricity. Renewable energy technologies can help countries meet their policy goals for secure, reliable, and affordable energy to expand electricity access and promote development. Among various micro hydropower plants, gravitational water vortex power plants are emerging because of their simple installation, reduced setup time, and minimal maintenance cost. There is a significant opportunity for the integrated development of water multi-purposes projects that can allow recovering this "sleeping" hydro potential. Even small irrigation or drinking water infrastructures can generate hydropower for the benefit of farmers cooperatives and Water User Associations, anywhere there.

Keywords: multi-village drinking water supply (MV-DWS), multi-use water system (MUWS), hydro-agricultural schemes (HAS), integrated development, micro-hydro power plants (mHPP), gravitational water vortex power plants

1. Introduction

A Multiple Use Water System (MU-WS) is designed to provide both domestic water and water intended for productive uses, such as irrigation (cash market gardening, cereals, fruit growing, hydroelectricity etc.), livestock breeding, or running a small business.

Hydropower is the largest renewable energy source, producing around 16% of the world's electricity and over four-fifths of the world's renewable electricity. Once implemented, hydropower plants usually require little maintenance and operation costs are low.

Nowadays and worldwide, the multi-purpose schemes operating in the water sector (irrigation and drinking water supply network) equipped with small hydropower plants are limited. There is, however, an exception in Switzerland where the Swiss Federal Office of Energy has produced guidelines for installing Small Hydro Power (SHP) plants (SFOE, 1996) to better promote energy recovery within water networks.

The potential for the development of multi-purpose pico and micro-hydroelectric power plants is very high in Africa and even Worldwide both for rural electrification and for Multiple Use Water Systems (MUWS) intended to provide both water for the integrated development of irrigation and hydroelectricity energy or mini–Drinking Water Supply (DWS) network coupled with a pico or micro or small hydroelectric power plant.

There is a huge potential for energy recovery through the already existing and future small, medium, and large hydro-agricultural schemes (HAS) which offer a great opportunity to (i) diversify sources of income (energy supply contract with mini-rice mills for rice husking in the valley), (ii) reduce the dependence of irrigation networks on power cuts, (iii) diversify production sectors and chains (electric drying of agricultural products, recharging of agricultural tractor batteries essential to ensure respecting the crop calendar in irrigated rice areas, etc.); (iv) power emergency generators (PEG) which are generally not provided at the pumping stations of hydro-agricultural schemes (HAS); (v) supply energy to the Health Centers and the offices of Cooperatives and Water User Associations (WUA) as well as those of the advice centers affiliated to them; (vi) reduce pumping cost and increase net income for the Cooperatives and Water User Associations.

2. Methodology for identification of small hydro power plants (SHPP) alongside irrigation schemes

The approach consists of identifying, from upstream to downstream, the installation points of pico and mini hydroelectric power plants at the level of:

i. The dam or the deadhead of the main canal (CP);

ii. Along the main canals by analyzing the longitudinal profiles in order to identify the singularities and the control sections of the water lines (hydraulic waterfall structures, water level regulation structures which are in principle regularly distributed every 300 to 500 meters or even 1000 meters, water intake and diversion works from the main canals to secondary canals, changes in slope along the main canals, etc.);

iii. Along secondary canals at the level of singularities, limits between hydraulic districts, intake structures, etc.

This identification will be made during the Project Preliminary Studies (PPS) of the Hydro-Agricultural Scheme or the Drinking Water Supply Network (DWS) and

DOI: http://dx.doi.org/10.5772/intechopen.1007635

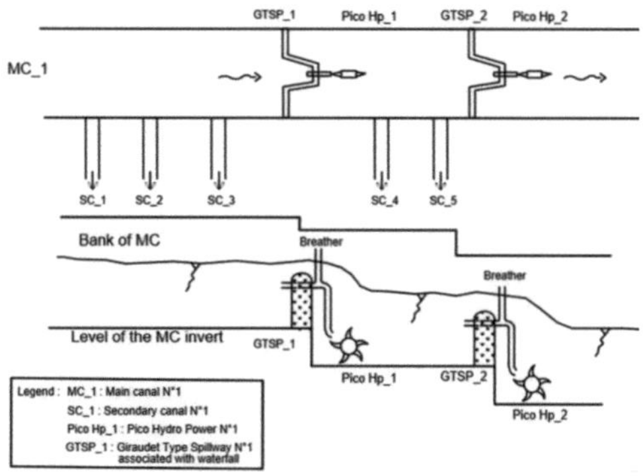

Figure 1.
Installation points of pico or micro-hydro power plant.

will be the subject of discussions by a "Task Force" at the same time as the debate on the choice of development options.

The option chosen for the integrated development of a multi-use water system for irrigation, Small Hydro Power Plants (SHPP) and drinking water supply will be the subject of a Detailed Feasibility Study (DFS) in the Detailed Preliminary Project phase.

Figure 1 illustrates a Main Canal (MC) along with a few Giraudet Type Spillways (GTSp) regulation structures associated with waterfalls that are regularly positioned to ensure a reliable water level within the MC. Thus, allowing installing low–waterfall pico/micro-Hydro – Power Plants (HPP) just behind the weir and cumulatively a proper feeding of Secondary Canals (SC) may be afforded. At the tail of the MC, it can be suitable to have Rectangular Type Spillways (RTSp), where the discharge in the MC is getting reduced, to water outlets feeding the SC upstream.

3. General conception of irrigated perimeters (IP) in Africa (particularly West and Central Africa)

For the design of irrigated areas, it is appropriate to adopt an integrated approach by considering the design of the layout plan for the irrigated or irrigable areas and the longitudinal profiles of the main, secondary and tertiary canals to dominate, at the same time, the land to be irrigated and tame the potential energies along the longitudinal profiles of the channels going from upstream to downstream.

3.1 Principles of operation of canals

3.1.1 Operating principle of irrigation networks in West and Central Africa

In certain irrigated areas in Africa, canals operate with upstream control. A determined flow rate is delivered upstream following a pre-established schedule. This flow must be taken downstream by diversion toward the hydraulic districts or sub-districts. The outlets of the downstream districts should be opened before delivering the upstream flow and closed after closing the upstream outlet. Water management must consider the response time of irrigation networks.

A reach of the main canal operates under downstream control and serves as a pump regulation reach.

This regulation is imposed by the fact that the flow rate supplied by the pumps cannot be strictly equal to the flow rate required in the network. In fact, the required flow rate is an integer multiple of the water flow (20 to 35 l/s for rice cultivation for example) while the pump flow rate is a multiple of the unit flow rate of the pumps (200 to 300 l/s for the hydro-agricultural rice developments of Bonfeba, Diomana and Dessa in Niger).

The regulation principle is as follows:

- The overall flow rate from the pumping station is greater than the flow rate required in the irrigation network.
- One of the pumps works intermittently.
- The excess volume delivered by this pump is stored in the canal reach.

This volume of the reserve, thus constituted, is delimited by a low level relative to the flow line of the maximum flow and a high level generally located 15 or 20 cm above,

When the high level is reached, the pump stops (automatically or manually). The flow rate of the pumps is then lower than the required flow rate and emptying the reserve completes the flow rate of the pumps in operation. The required flow rate is thus provided.

The size of the reserve is done as follows:

The flow rate Qr of the irrigation network is between the flow rate of "N" pumps of unit flow rate q and the flow rate of "(N-1)" pumps:

$$(N-1)\times q < QR < N\times q \tag{1}$$

$$Qr = \mathrm{N}\times \mathrm{q} - \mathrm{a}\times \mathrm{q} \tag{2}$$

One of the "N" pumps operates intermittently; it is started to fill the reserve and it is stopped during emptying. The duration of the cycle breaks down as follows:

Filling time:

$$T1 = \frac{V}{N\times q - Qr} = \frac{V}{q}\times\frac{1}{a} \tag{3}$$

with n pumps in operation.

Emptying time or drain time:

$$T2 = \frac{V}{Qr-(n-1)\times q} = \frac{V}{q}\times\frac{1}{1-a}, \tag{4}$$

with (n-1) pumps in operation.

Total cycle duration:

$$T = T1 + T2 = \frac{V}{q}\times\left(\frac{1}{a}+\frac{1}{1-a}\right) \tag{5}$$

The minimum time which separates a stop from a start is equal to:

$$\frac{V}{q}, \tag{6}$$

which corresponds to $a = 0$ considering Eq. (4).

The duration of the complete cycle $T = T1 + T2$ as per Eq. (5), separating two starts or two stops passes through a minimum for $a = \frac{1}{2}$, which gives:

$$T = \frac{4 \times V}{q}, \tag{7}$$

The minimum volume of the reserve is chosen according to the unit flow rate of the pumps and the operating constraints of the pumping groups.

A supply channel allows water from the river (mostly perennial) to be transported to the pumping station, sized to meet the water needs of a hydro-agricultural scheme (often rice growing) of surface area S (e.g., 800 ha), using n + 1 pumps including an emergency pump. For areas irrigated by pumping (i.e., cases of Niger, Rwanda, Burundi, Burkina Faso, Cameroon), the pumps lift the water into a dead head which is regulated by command from downstream: the water line in the dead head of the main canal fluctuates between a maximum level and a minimum level thus providing a regulation volume. For example, for an irrigated area (IA) of 800 ha, the flow rate required at the head of the network is 2400 l/s (at a rate of 3 l/s/ha for rice growing). The station will include six pumps of 500 l/s each, including five operating pumps and one emergency pump. The station will operate with four pumps continuously for 16 hours per day and one intermittent pump which will be used to regulate the dead head of the main canal using an automatic bulb.

The main canals take water from the dead head of the main canal or Supra Main Canal (SCP) through intake structures (generally mask modules) and follow the contour lines that overhang/dominate the basins and swamps.

The mask modules supply water to the secondary channels (CS) which are derived from the main canals (MC) by crossing the contour curves. They in turn supply tertiary canals (CT), which follow the contours, using modulated outlets (modulated mask/taps).

The tertiary canals (TCs) supply water to planned[1] plots *via* plot intakes (and sometimes siphons).

4. Current situation of the development of micro-hydroelectric power stations alongside modern water points

Integrated development of micro-hydroelectric power stations in parallel with modern water points is very timid. Some East African countries such as Rwanda, Burundi, Tanzania and Uganda have developed micro-hydroelectric plants for the benefit of rural populations, particularly for rural electrification.

In Rwanda, for example, more than 23 micropower plants have been tested and distributed in suitable areas. An informal sub-sector for pico-hydro (hydro plants below 50 KW) already exists in Rwanda. There are at least 10 micros enterprises or individuals active in the manufacturing and installation of pico-hydro turbines.

[1] Sometimes by laser.

4.1 Integration of micro hydroelectric plants in irrigated areas

Specifically, the state of the art in terms of integrated development of micro/pico power plants at the level of irrigated areas and Drinking Water Supply (DWS) networks is also very weak.

There are currently only a few initiatives to conceptualize the integrated development of pico and micropower plants, among which we can cite:

i. Gravitational Water Vortex 15 kW Mini Hydro Turbine prepared by Dr. Gurmit Singh, January 2019;

ii. Integration of Small Hydro Turbines into Existing Water Infrastructures; Aline Choulot[2] (Mhylab), Vincent Denis (Mhylab), and Petras Punys[3];

4.2 Integration of micro-hydroelectric power plants into modern drinking water supply (DWS) networks

Experiences in the integrated development of micro or pico hydroelectric power plants on the fringes of the Drinking Water Supply (DWS) network are almost non-existent even at the global level (not only in Africa).

5. What types of micro/pico plants in irrigated areas (IA) and in drinking water supply (DWS) networks?

5.1 Definitions

A small hydropower plant (micro power plant) is an energy production installation based on the use of hydraulic power and whose power is less than 300 kW.

Hydraulic energy can come from a river, a spring, a drinking water supply network, a wastewater or drainage network, industrial processes in which the pressure of a liquid is destroyed in a regulating device, valve or other equipment. The production of energy by a hydraulic power plant – whether large or small – depends on the water flow and the head (pressure difference) available.

As in Ref. [1], the principle of a Small Hydraulic Power Plant is given in **Figures 2** and **3**.

Hydroelectric power requires a watercourse or reservoir. The power plant exploits the potential energy created by the height difference of the waterfall. It consists of a reservoir or a stream, a waterfall characterized by its height. Hydraulic energy is converted into electricity by passing water through a turbine connected to an alternator. The power of the plant depends on the water flow.

The formula for estimating the available hydraulic power is:

$$P = 9.81 \times Q \times H \tag{8}$$

[2] Switzerland Mini-Hydraulics Laboratory (Mhylab).

[3] Water and Land Management Faculty, Lithuanian University of Agriculture.

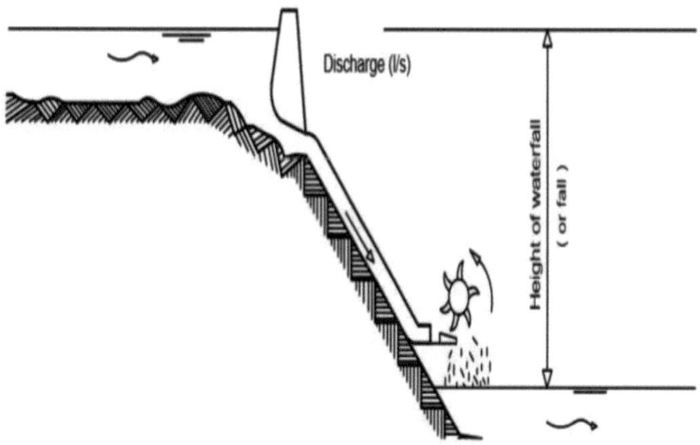

Figure 2.
Principle of small hydraulic power plant.

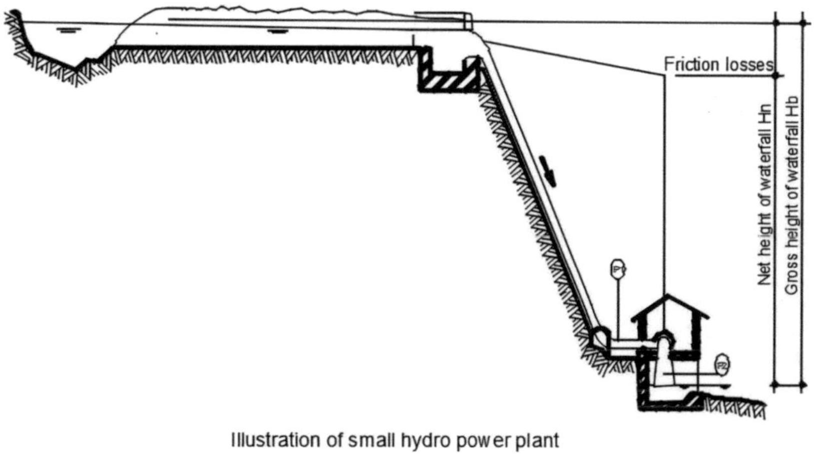

Figure 3.
Height of waterfall.

With: P, the hydraulic power in Watts (W). Q, the flow rate in liters per second (l/s). H, the height of fall in meters (m).

As per Ref. [1], the height of the waterfall is illustrated in **Figure 2**.

To calculate the truly usable power, it is necessary to multiply the hydraulic power (determined above) by (η) the efficiency of the turbine and generator installation, generally between 0.6 and 0.9.

Figures 2 and **3** are for illustrative purposes only. However, the analysis of Eq. (6) reveals that regarding the overall efficiency of the pico/micro–Hydroelectric Power Plant to be considered in the design, the literature generally cites an efficiency between 60 and 90%. However, it is advisable to be more cautious and to adopt 50% accordingly to consider the losses in output from upstream to downstream and to anticipate the wear and aging of the mechanical components when the plant has reached 10 years of operation.

From two meters (even less according to the discharge), a fall with sufficient flow is considered usable. Furthermore, heights of 500 meters (pressure 50 bars) or more can produce an interesting quantity of energy, even if the flow rate is low (5 l/sec).

There is a consensus on the classification of hydroelectric power stations according to power:

i. Classification of hydroelectric power plants according to the International Union of Producers and Distributors of Electrical Energy (IUPDEE):

As in Ref. [2], the classification established by the International Union of Producers and Distributors of Electrical Energy (IUPDEE) highlights several types of power plants among which four (4) types of small power plants:

- The pico-central for a power less than 20 Kw;
- The micropower plant for a power between 20 and 500 Kw;
- The mini-power plant for a power between 500 and 2000 Kw;
- The small power plant for a power between 2000 and 10,000 Kw;
- Large hydroelectric power plants/Full Scale Hydro FSH: Power greater than 10 MW.

ii. Classification of hydroelectric power plants, case of Rwanda as in Ref. [3]:

- o Pico hydraulic power stations: Power P less than 50 Kw;
- o Micro-hydraulic power plants (Micro Hydro Scheme MHS): P less than 2.5 MW;
- o Small Hydroelectric Power Plant/Mini Hydro Scheme: P 2.5 to 10 MW;
- o Large hydroelectric power plants/Full Scale Hydro FSH: P greater than 10 MW.

iii. Classification of Adams Harvey hydroelectric power stations (USA) as in Ref. [4]:

- o Pico hydraulic power plants: P less than 200 Kw (Rwanda);
- o Micro-hydraulic power plants (Micro Hydro Scheme MHS/Small Hydropower Plant/SHPP): P 200 to 300 KW (Adams Harvey);
- o Small Hydroelectric Power Plant/Mini Hydro Scheme: P 300 KW to 10 MW (Adams Harvey);
- o Large hydroelectric power plants/Full Scale Hydro FSH: P greater than 10 MW.

5.2 Types of micro/pico plants for irrigated perimeters (IP)

5.2.1 General case of low pressure installations along a watercourse or on a diversion canal

The most important structure is the dam or water intake, most often built-in concrete. Its function is to divert the necessary flow directly to the turbine or into a diversion channel while allowing floods to pass. The power plant is either integrated directly into the dam or placed at the end of a canal. Generally, there is no forced conduct, or it remains very short. The falls are between 2 and 20 meters and the pressure in the turbine is low (0.2 to 2 bars).

5.2.2 Case of the gravitational water vortex power plant (GWVPP)

It may be of great interest to use Gravitational water vortex turbines which are a micro-hydroelectric technology used at low heads (0.75–3 m). They create a swirling vortex that is used to drive an impeller. They were rejected primarily because of their inability to effectively handle the high heads (34.5–45 ft) and inlet and outlet piping particularly to this task.

The Gravitational Water Vortex Power Plant (GWVPP) is an example of a cost-efficient micro hydropower turbine that can be implemented to deliver local renewable and reliable energy without any impact.

The turbine is specifically designed to efficiently use a large flow with low pressure allowing for bigger tolerances in the total construction and reaching all fish-friendly parameters. Sediment and small debris can pass through without the need for additional structures. Because of the relatively small size, prefabricated structures can be used, decreasing the local construction complexity. All this lowers the civil works and installation costs significantly. The cost of energy is often inferior to 0,03 USD/kWh.

An example of cost sharing for several components of a mini-Hydro Power (HP) as per Ref. [5] shows that the civil works and the turbine generator represent 40 and 30% respectively while the control equipment can be roughly estimated as 22% of the total cost. The management cost will account for 8% of the cost of the mini-HP.

In Ref. [6], the Hydro Power Plant for "canal/conduit projects below 1 MW" shows similar breakdown cost percentages declined as follows: Civil works 41%; electro–mechanical equipment 37.1%; electrical infrastructure 10.3%; engineering and construction management 11.6%.

Likewise, Ref. [7] highlights the following breakdown costs percentages: Hydrotechnical constructions 55%; turbines 30%; Buildings 5%; electrical equipment 10%.

As a result, the distribution of costs for the components of a mini hydroelectric power plant indicated in Refs. [6–8] are in the same order of magnitude.

Regarding costs, Ref. [7], it is indicated a total cost of approximately 5500 US dollars per micro power plant. For an irrigated area of 500 Ha for example, it is reasonable to have 10 to 15 micro power plants of 6000 USD each, that is, an investment of around 60,000 to 90,000 USD, which amounts to a specific cost of 120 to 180 USD per hectare developed.

Considering an average cost per hectare developed of around 15,000 USD, it turns out that the overall cost of building 10 to 15 micro-hydroelectric power plants hardly represent more than 0.8 to 1.2% of the cost of the hydro-agricultural development, which is acceptable.

It is therefore appropriate to consider this distribution of costs in maintenance and upkeep charges as well as in the evaluation of the cost of water. A global plan for the operation and maintenance of multiple use water infrastructures for micro and small hydroelectricity would benefit from being systematized.

An overall assessment of the financial and economic profitability will be carried out.

Finally, supply the databases and knowledge-sharing platforms which can serve as support for the Public Private Partnership (PPP).

As described in Ref. [7] The Gravitational Water Vortex (GWV) Power Plant is a horizontal form of the hydroelectric dam. The benefits of using an artificially induced vortex above gravity-accelerated water increase efficiency, decrease cost, and not only lower the negative impact on the environment but also increase sustainability and health of the river.

As illustrated in **Figures 4** and **5**, River water is channeled at the bank of the river and conveyed to a circulation tank. This circulation tank has a circular orifice at its base. The combination of localized low pressure at the orifice and the induced circulation at the tangential entry influences strong vortex flow. Potential energy is entirely converted to rotational kinetic energy at the vortex core. This is then extracted by means of a vertical-axis turbine. Water is then returned to the river through the tail race.

In the vortex power plant, water is introduced into a circular basin tangentially, creating a free vortex and energy is extracted from the free vortex by a turbine.

The turbine, which is located at the top of the vortex, is turned by the rotary motion of water in the vortex.

This type of power plant can be installed at rivers or streams or at the head of main or secondary canals as the hydraulic head requirement is as low as 1 m. Water may be fed to the vortex turbine through an open channel or, in some cases, through a closed conduit or pipe, which is more suitable for smaller systems.

5.2.3 Power output and efficiency

The maximum hydraulic power output of the turbine is given by:

Figure 4.
Illustration of gravitational water vortex power plant.

Figure 5.
Illustration of gravitational water vortex power plant.

$$P = \eta \times \rho \times g \times Q \times H \tag{9}$$

where

P = Usable Power in watt; H_v = Height of vortex (m); Q = Flow rate (m^3/s); g = Gravitational constant; ρ = Water density (kg/m^3) = ç = the efficiency of the hydro power plant installation, generally between 0.6 and 0.9. Recommended long run efficiency 0.5.

It derives from the equation above that a maximum hydraulic output of 1 KW can be obtained with a flow rate of only 0,1 m^3/s and a head of only 0,6 m (applicable to main and secondary irrigation canals). The simple design of the turbine makes it easy to downscale to small sizes. The power output is influenced by the height of the vortex and the water flow rate Q.

The construction cost of a GWV Power Plant is affordable (5520 USD for a 15 kW Power Plant).

Table 1 shows that the gravitational water vortex (GWV) power plant is not the only low-head power plant option that can be used in multi-use water points for hydropower development. There is other lower-cost alternatives that can be used for low-head pico and micro power plants such as (i) Kaplan Propeller, Kaplan – Bulbe, Reversible Turbine Pump as turbine PAT) for irrigated areas and (ii) Francis, Kaplan Propeller for Rural and urban Drinking Water Supply network (DWS).

All low-head technologies can be associated with fall structures on main or secondary canals and water intake and diversion structures from primary canals to secondary ones.

Types of water points	Height (m)	Flow rates (m^3/s)	Type of hydroelectric power plant	Turbine type	Observations
Large irrigated areas from 1001 to 15,000 Ha	2.5–7	3–45	Low fall height (low-head)	• Kaplan • Propeller • Kaplan – Bulbe • Reversible Turbine (Pump as turbine PAT)	Reversible Turbine: To turbine the drainage water and discharge it toward the river at low flow. During High flow, the installation will be used to pump water from the main drain into the river.
Medium irrigated areas from 501 to 1000 Ha	1.5–2.5	1.5–3	Low fall height (low-head)	• Kaplan • Propeller • Kaplan - Bulbe • GWV	
Small irrigated areas of 100 to 500 Ha	0.5–1.5	0.3–1.5	Low fall height (low-head)	• Kaplan • Kaplan - Bulbe • GWV	
Mini Rural DWS networks (H 10 m; low flow)	10	0.14–0.28	Low fall height (low-head)	• Kaplan • Propeller	
Urban DWS networks (H 20–30 m; high flow)	20–30	1–2	Low fall height (low-head)	• Francis • Kaplan • Propeller	

Table 1.
Types of micro/pico hydroelectric power plants for irrigated perimeters (IP) and drinking water supply (DWS) networks.

5.3 Types of micro/pico hydroelectric power plants for drinking water supply (DWS) networks

5.3.1 Medium and high-pressure installations on rivers, mountain springs, drinking water networks

The mini–Drinking Water Supply (DWS) networks installed in rural areas in Africa in general and in particular in East, West, Central Africa and in the Large Islands such as Madagascar are of relatively modest size with heights of around 10 m and flow rates varying from 50 to 100 m^3/h (i.e., 14 to 28 l/s).

For this type of infrastructure, it is possible to install Kaplan (low-head) turbines with a vertical axis. Pelton turbines are more reserved for "high head" hydraulic power plants. If the flow variation is controllable (below 10%), Franci's turbines can be used.

"Cross – flow" type turbines can also be used but their efficiency is lower than that of Kaplan and Francis turbines (under equal conditions).

The urban Drinking Water Supply (DWS) networks are larger. They use water towers whose height can reach 20 to 30 m (rarely more) with peak flow rates of around 100 to 1000 l/s (or even 1 to 2 m^3/s). Kaplan and Francis turbines may be suitable.

Table 1 resumes the type of turbines to be used for large (1001 to 15,000 Ha), medium (501 to 1000 Ha) and small-scale (100 to 500 Ha) irrigation schemes on one hand and those to be implemented for Mini Rural DWS networks and Urban DWS networks on the other hand.

6. Approach for smart integration of micro and pico hydroelectric plants in irrigated areas

6.1 Different energy recovery possibilities at the level of the dead head of main canal or the supra main canal (SMC)

- Where to place the micro/pico hydroelectric power stations:
    - At the level of the feed channel connecting the river to the pumping station (ideal site for a GWV with a significant margin of progress in the recovery of neglected or dormant energies);
    - At the level of the Dead Head (DH) of the Main Canal (MC) or supra main channel (SMC): it is possible to place a micro-central at the start of each main channel (CP) using the regulation water slice and the water blade in the Dead Head of Main Canal (DH);
    - There will be as many Main Canals (MC) as there are micro-hydroelectric plants whose power lines will be connected (mostly in parallel) before being connected to à transformer or a storage inverter.
- For what uses:
    - To power emergency generators (GE) which are generally not provided at the pumping stations of hydro-agricultural schemes (HAS);
    - To compensate for power outages which have become frequent in most Central and West African countries;
    - To recharge the batteries of tractors that are used in Central Africa (case of the Company for Modernization and Expansion of Rice Farming of YAGOUA "SEMRY" in the Far North of Cameroon) and which can also be popularized in other countries of West Africa to avoid delays in plowing which is a main determinant for respecting the crop calendar to avoid the overlap of dry season and rainy season campaigns.
    - To promote and install mini-rice mills (1 Tons/H for irrigated perimeters (IP) of average size 1000 Ha; even 4 to 5 Tons/H for large IP of the order of several thousands of hectares);
    - To supply energy to the Health Centers and the offices of cooperatives and WUAs as well as those of the advice centers affiliated with them.

6.2 Different energy recovery possibilities at the level of the main channels/canals (MC)

- Where to place the micro/pico hydroelectric power stations
    - At the level of the main canals (CP): it is possible to place a micropower plant at the start of each main canal (CP) and at the level of the waterfalls

which must be the subject of particular attention during the plan layout and the longitudinal profile by exploiting the steep slopes of the land to create significant falls at the level of the concrete channels and the water level in the main channels (CP);

- o At the level of singularities (sudden changes in slopes or sections, etc.);
- o At the junction between the dead sections of the river (which collect water from the main drains) and the water level in the river at low flow;
- o Micro/pico hydroelectric power stations could, as far as possible, be placed at the level of the lateral safety spillways and at the level of the transversal spillways for regulating the water level which is generally provided at regular distances (every 300 to 500 meters for example);
- o It is also possible to place them at the level of inverted siphons which are passage structures under laterite or earth tracks/roads for crossing a ravine/a watercourse/a "Kori":
- o There will be as many secondary canals (CS) as there are micro-hydroelectric plants whose power lines will be connected (mostly in parallel) before being connected to a transformer or a storage inverter.

For what uses

- To power emergency generators (GE) which are generally not provided at the pumping stations of hydro-agricultural schemes (HAS);
- To compensate for power outages which have become frequent in most Central and West African countries;
- To recharge the batteries of tractors that are used in Central Africa (case of SEMRY in the Far North of Cameroon) and which can also be popularized in other West African countries to avoid delays in plowing which is a main determinant for respecting the crop calendar to avoid the overlap of dry season and wet season campaigns.
- To promote and install mini-rice mills (1 Tons/H for irrigated areas (IA) of average size 1000 Ha; even 4 to 5 Tons/H for large irrigated areas (IA) in the range of several thousand hectares);
- To supply energy to the Health Centers and the offices of Cooperatives and WUAs as well as those of the advice centers affiliated with them;
- To supply energy to agricultural product dehydrators for women's and youth groups and private business centers (on demand).

6.3 Different energy recovery possibilities at the secondary canal level (SC)

- Where to place the micro/pico hydroelectric power stations:

- o At the level of the secondary canals (CS): it is possible to place a micropower plant at the start of each secondary canal (CS) and at the level of the Waterfalls which must be the subject of particular attention during the plan layout and of the longitudinal profile by exploiting the steep slopes of the land to create significant falls at the level of the concrete channels and the water level in the secondary channels (CS);

- o Micro/pico-central units could, as far as possible, be placed at the level of the lateral safety spillways and at the level of the transversal spillways for regulating the water level which is generally provided at a regular distance (every 300 meters for example);

- o It is also possible to place them at the level of inverted siphons which are passage structures under tracks/roads for crossing a ravine, a watercourse, a Kori:

- o The number of micro-hydroelectric plants will be reasoned on a case-by-case basis depending on the configuration of the land and the torque (flow, water level) in the secondary canals (CS). The electrical lines can, if necessary, be connected (mostly in parallel) before being connected to a transformer or a storage inverter.

For what uses

- To power emergency generators (GE) which are generally not provided at the pumping stations of hydro-agricultural developments (AHA);

- To compensate for power outages which have become frequent in most Central and West African countries;

- To recharge the batteries of tractors that are used in Central Africa (case of SEMRY in the Far North of Cameroon) and which can also be popularized in other West African countries to avoid delays in plowing which is a main determinant for respecting the crop calendar to avoid the overlap of dry season and rainy season campaigns.

- To promote and install mini-rice mills (from 0.5 to 1 Ton/H for irrigated perimeters (IP) of average size 1000 Ha; even 4 to 5 Tons/H for large irrigated perimeters (IP) in the range of several thousand hectares);

- To supply energy to the Health Centers and the offices of Cooperatives and WUAs as well as those of the advice centers affiliated with them;

- To supply energy to dehydrators of agricultural products for women's groups and young people and private business centers such as cell phone charging centers or storage of agricultural products (on demand);

6.4 Different energy recovery possibilities at the tertiary canal (TC) level

- Where to place the micro/pico hydroelectric power stations:

    - o The number of micro hydroelectric plants will be reasoned on a case-by-case basis depending on the configuration of the land and the couple (flow, water level) in the secondary canals (CS). The electrical lines can, if necessary, be connected (mostly in parallel) then connected to a storage inverter or to a transformer.

For what uses:

- To supply energy to dehydrators of agricultural products for women's groups and young people and private business centers such as cell phone charging centers or storage of agricultural products (on demand).

7. Approach for smart integration of micro and pico hydroelectric power stations in modern drinking water supply (DWS) networks

7.1 Different energy recovery possibilities in the case of spring at altitude

A simple drinking water network can comprise the following components: (i) a spring at altitude, (ii) a forebay, (iii) a penstock, (iv) a reservoir, (v) a water supply network.

From the elevation of the sources, very often, there can be an excess of pressure in the networks to recover. There may be a great opportunity to replace the pressure breakers, which are usually intended to break the excess pressure, with turbines so as to generate electricity.

- Where to place the micro/pico hydroelectric power stations:

    - On a reservoir: Water passes through the turbine before being accumulated in a reservoir. This method is the most flexible, as it permits disconnection of the turbine operation from the water supply network to guarantee at any time the primary function of the existing infrastructure.

    - Within the supply network: Water passes through the turbine and continues its path through the pipe. Therefore, a pressure defined by the network requirements has to be maintained at the turbine outlet, which can be allowed by reaction turbines and counter-pressure Pelton;

    - At the foot of the water towers: it is possible to place a micropower plant at the start of the service pipe by valuing the altitude of the tower base and the water height in the castle/tower;

For what uses

- To supply energy to the Health Centers and WUA offices as well as those of the advice centers affiliated with them;

- To supply energy to private business centers such as mobile phone charging centers or rural shopping centers (on demand).

7.2 Different energy recovery possibilities in case of underground pumped water

- Where to place the micro/pico hydroelectric power stations:

DOI: http://dx.doi.org/10.5772/intechopen.1007635

- When the drinking water source is underground and must be pumped to the reservoir, turbine integration will not be easy.

- At the foot of the water towers: using a reaction turbine or Pelton turbine, it may be possible to settle a micropower plant at the start of the service pipe taking advantage of the pressure head between the elevation of the tower base and the water height in the tower;

- For what uses

  - To power emergency generators (GE) generally provided at the pumping stations of Drinking Water Supply (DWS) networks;

  - To compensate for power outages provided by the national electricity network or maintenance inadequacies often noted at the level of generators and small photovoltaic solar units;

  - To supply energy to the Health Centers and WUA offices as well as those of the advice centers affiliated with them;

  - To supply energy to private business centers such as mobile phone charging centers or rural shopping centers (on demand).

8. Application to the irrigated perimeter (IP) of Konni (2452 Ha downstream of Mozagué and Zongo dams in Niger); IP of SEMRY1 (5500 Ha supplied by pumping stations in YAGOUA Cameroon) and SEMRY2 (6000 HA downstream of the MAGA dam in Cameroon)

Table 2 resumes the estimation of Power generated by the pico or micro hydroelectric plant P (W), the monthly energy generated in Kilowatt Hours, the Gross Income generated by energy production and electricity consumption equivalent households for Konni Irrigated scheme in Niger (2452 Ha).

As demonstrated in **Table 2**, total energy recovery at the level of all primary canals of the Konni Irrigated scheme using Gravitational Water Vortex Power Plants at the level of CA; CP_A, CP_B, CP_C, CP_D, can be a good opportunity for the Water User Association to Harness 183,145 KwH per year. This can result in a substantial Gros Revenue generated of 17,948 USD per year. These pico power plants can allow an electricity consumption equivalent household of 2544.

Table 3 illustrates the estimation of Power generated by the pico or micro hydroelectric plant P (W), the monthly energy generated in Kilowatt Hours, the Gross Income generated by energy production and electricity consumption equivalent households for SEMRY2 irrigation scheme in Cameroon (6000 HA downstream of the MAGA dam).

s illustrated in **Table 3**, total energy recovery at the level of all primary canals of Konni Irrigated scheme using Gravitational Water Vortex Power Plants at the level of MC of Pouss, Maga Est, Maga West, Guirvidic, can be a good opportunity for the Water User Association to Harness 59,770 KwH per year. This can result in a substantial Gros Revenu generated of 105,359 USD per year. This pico power plants can allow an electricity consumption equivalent household of 9962.

CH	N° Canal	Q (l/s)	H (m)	Did	Nidm	P (W)	Em (KwH)	Ean (KwH)	Prix KwH	GIe (USD)	EcEqH
CA	FC	8090	1,2	8	12	66,664,8	6399,8	76,797,9	0,098	7526,2	1067
MC	MC_A	2130	2	8	12	29,253,4	2808,3	33,699,9	0,098	3302,6	468
MC	MC_B	3030	2	8	12	41,614	3994,9	47,939,4	0,098	4698,1	666
MC	MC_C	1060	1,53	8	12	11,136,9	1069,1	12,829,7	0,098	1257,3	178
MC	MC_D	910	1,65	8	12	10,310,8	989,8	11,878,0	0,098	1164,0	165
Total ER at the level of primary canals						158,980	15,262,1	183,144,9		17,948,2	2544

NB: HC: Channel Hierarchy; CA: Feeding channel; MC: Main Canal; N° Canal: Channel Designation; Did: Daily irrigation duration; Q: Discharge (l/s); H: water Height (m); Nidm: Number of irrigation days per month; P (W): Power generated by the pico or micro hydroelectric plant; Em (KwH): Monthly energy generated in Kilowatt Hours; GIe: Gross Income generated by energy production; EcEqH: Electricity consumption equivalent households; ER: energy recovery.

Table 2.
Konni Irrigated scheme (2452 Ha downstream Mozagué and Zongo dams (Niger).

DOI: http://dx.doi.org/10.5772/intechopen.1007635

HC	N° Canal	Q (l/s)	H (m)	Dj	Njm	P (W)	Em (KwH)	Ean (KwH)	Prix KwH	Re (USD)	EcEh
MC	MC Pouss	8000	2,8	16	4	153,820,8	9844,5	118,134,4	0,15	17,353,2	1641
MC	MC Maga Est	16,000	2,8	16	4	307,641,6	19,689,1	236,268,7	0,15	34,706,4	3282
MC	MC Maga West	16,000	3,2	16	4	351,590,4	22,501,8	270,021,4	0,15	39,664,5	3750
MC	MC Guirvidig	8000	2,2	16	4	120,859,2	7735,0	92,819,9	0,15	13,634.7	1289
Total at the level of primary canals						933,912	59,770,4	717,244,4		105,358,9	9962

Table 3.
IP of SEMRY2 (6000 HA downstream of the MAGA dam in Cameroon).

Dpt	Water Tank	Q (l/s)	H (m)	Dwsd	Njm	P (W)	Em (KwH)	Ean (KwH)	Prix KwH	Re (USD)	EcEh
BOUR	191	2,8	9	10	30	191	57	687	0,227	156,1	10
YEGUIL	248	2,8	12	10	30	248	74	893	0,227	202,9	12
Total at the level of Primary Canals						439	132	1579		359,0	22

NB: Dpt: Department; FC: Feeding Channel; MC: Main Canal; N° Canal: Channel Designation; Dwsd: Daily water supply duration; Q: Discharge (l/s); H: water Height (m); Nidm: Number of irrigation days per month; P (W): Power generated by the pico or micro hydroelectric plant; Em (KwH): Monthly energy generated in Kilowatt Hours; GIe: Gross Income generated by energy production; EcEqH: Electricity consumption equivalent households; ER: energy recovery.

Table 4.
BOUR and YEGUIL mini–drinking water supply (DWS) (KANEM region).

9. Applications to the mini: Drinking water supply (DWS) networks of BOUR and YEGUIL (KANEM region in CHAD)

Table 4 highlights the estimation of Power generated by the pico or micro hydroelectric plant P (W), the monthly energy generated in Kilowatt Hours, the Gross Income generated by energy production and electricity consumption equivalent households for BOUR and YEGUIL mini–Drinking Water Supply (DWS) (KANEM Region).

As shown in **Table 4**, total energy recovery at the level of all primary canals of the Konni Irrigated scheme using Gravitational Water Vortex Power Plants at the level of BOUR and YEGUIL drinking water supply networks can be a good opportunity for the Water User Associations to Harness 1579 KwH per year. If though the Revenue of 359 USD generated per year seems to be very low. These pico power plants can allow charging telephones for instance and lighting health centers. Moreover, this setting represents an electricity consumption equivalent household of 22.

10. Development potential and advocacy for the integration of micro and pico hydroelectric plants in irrigated areas (PI) and drinking water supply (DWS) networks in west and Central Africa

The potential for the development of multi-purpose pico and micro-hydroelectric power plants is very high in Africa and even Worldwide both for rural electrification and for Multiple Use Water Systems (SEUM) which are intended to provide both water for the integrated development of irrigation and hydroelectricity energy or mini–Drinking Water Supply (DWS) network coupled with a pico or micro-hydro-electric power plant.

For instance, in Rwanda, a minimum available potential of 168 pico-hydro sites has already been identified, but the actual potential is substantially higher because most of very small sites have not been recorded yet. In addition, climate change is projected to positively affect hydro resources in Rwanda.

Already existing and future small, medium and large hydro-agricultural schemes (HAS) offer a great opportunity to (i) diversify sources of income (energy supply contract with mini-rice mills for rice husking in the valley), (ii) reduce the dependence on irrigation networks on power cuts, (iii) diversify production sectors and chains (electric drying of agricultural products, recharging of agricultural tractor batteries essential to ensure respecting the crop calendar in irrigated rice areas, etc.).

11. Conclusion and recommendations

Renewable energy technologies can help countries meet their policy goals for secure, reliable and affordable energy to expand electricity access and promote development. Among various micro hydropower plants, gravitational water vortex power plants are emerging because of their simple installation, reduced setup time, and minimal maintenance cost.

Existing examples show that the electricity provided by even very small installations has a phenomenal impact on remote areas, which are usually synonym with deprived areas. Locally manufactured pico-turbines are significantly cheaper than imported ones and pico-hydro systems have much lower investment costs than PV systems or even micro-hydro plants and much lower operating costs than diesel. In addition, a pico-hydro installation can be completed within 6 months only. This presents an opportunity to obtain electricity quickly and affordable for the poorest segment of the rural population.

There is great potential and opportunity for the development of pico and micro hydropower in the existing irrigation schemes. An integrated approach should be also envisaged for future Multi-Use Water Service development along with the future hydro-agricultural infrastructures and drinking water networks.

Therefore, it can be recommended to launch a support strategy in order to test and or scale up the design and implementation of a pico and micro-hydro sub-sector program.

Acknowledgements

This paper is based on the literature review and the experience of the author in the water sector including irrigation, drinking water supply and hydro power.

This work has been self-supported by the author. The author would like also to thank the reviewers for their valuable comments and my beloved son M. Adam Ahmed Abdou (Renewable Energy Engineer) for his support for the drawings of **Figures 1–3** and his participation in the review of the chapter.

Conflict of interest

Illustrations copyright:

- Concerning **Figures 2** and **3**, the authors have reproduced by redrawing on autocad, the illustrations on pages 9 and 18 of the Practical Guide for the realization of Small hydropower plants [1]. There are no conflicts of interest because in the copyright of September 1992, the following mention is specifically made; "any reproduction of extracts is authorized with indication of the source".

- Regarding **Figures 4** and **5**, the authors used two photos from [7]. There is no conflict of interest because it is a white paper (available on the internet). This does not require permission, but the author has been cited.

Author details

Ahmed Moulaye Abdou[1*] and Adam Ahmed Abdou[2]

1 Single Project Implementation Unit RAB/MINAGRIE, Kigali, Rwanda

2 International Foundation for Water and Environment, Ouagadougou, Burkina Faso

*Address all correspondence to: abdoumoulaye@yahoo.fr

References

[1] Chapallaz J-M, Eichenberger P. Guide pratique pour la réalisation de Petites centrales hydrauliques. Bureau d'ingénieurs, programme d'action énergies renouvelables (PACER) de l'Office fédéral des questions conjoncturelles; 1992. ISBN 3-905232-20-0 1992, 96 pages, Copyright © 1992 Federal Office for Economic Affairs, 3003 Bern, September 1992. Reproduction of extracts authorized with indication of the source. Diffusion: French-speaking coordination of the "Construction and energy" action program EPFL-LESO, PO Box 12, 1015 Lausanne (Order number 724.244 f). Sainte-Croix. Layout and photocomposition: Consortium DAC / CITY COMP SA Lausanne and Morges

[2] Breuillot I. Guide méthodologique pour les projets d'hydroélectricité 2 ASDER. Pôle Collectivités, Association Savoyarde pour le Développement des Énergies Renouvelables; Chambéry; 2018, 48 pages

[3] PSP Hydro Project Rwanda. Support to the Pico-Hydro Sub-Sector Feasibility Analysis and Strategy. Kigali: PSP Hydro Team; 2010, 52 pages

[4] Harvey A, Brown A, Hettiarachi P, Inversin A. Micro Hydro Design Manual a Guide to Small – Scale Water Power Schemes (USA) as in Reference. London, UK: Intermediate Technology Publications; 1993

[5] Laghari JA, Mokhlis H, Bakar AHA, Mohammad H. A comprehensive overview of new designs in the hydraulic, electrical equipments and controllers of mini hydro power plants making it cost effective technology. Renewable and Sustainable Energy Reviews. 2013;**20**:279-293. Available from: https://www.sciencedirect.com/science/article/abs/pii/S1364032112006983?via%3Dihub www.elsevier.com/locate/rser

[6] O'Connor P, W, Zhang QFK, DeNeale ST, Chalise DR, Centurion E. Hydropower Baseline Cost Modeling. Oak Ridge National Laboratory. Springfield, VA: National Technical Information Service; 2015, 47 pages. Available from: http://www.ntis.gov/support/ordernowabout.htm

[7] Gurmit Singh, Gravitational Water Vortex 15 kW Mini Hydro Turbine [Internet]. 2019. Available from: https://5.imimg.com/data5/SM/YL/CC/SELLER-7846758/micro-hydro-turbine.pdf [Accessed: April, 21, 2024]

Chapter 6

Pumped Storage

Jamal El Mansour

Abstract

Pumped storage systems (PSS) is the largest worldwide battery system to store excess energy and manage the balance between electricity consumption and production. Using the Francis turbine as a turbine or pump makes the development of PSS feasible and economically accepted. Pumped storage is classified as low-, medium-, and high-head power plants. It needs two separate vertical reservoirs. In the case of using the sea as a lower reservoir, we reduce the construction cost and improve the system viability by an illimited reservoir. The coastline line areas are suitable for implementing these systems with very interesting altitudes (up to 400 m). The erosion due to the seawater depends on velocity distribution in the blade-to-blade section from the hub to the shroud section. For the pump, the erosion rate is as high as the trailing edge where the absolute velocity is high.

Keywords: energy transition, pumped storage, seawater, hydraulic pump, erosion

1. Introduction

Due to the evolution of industry (4.0 Generation), human behaviors, and citizens' habits, the need for electricity access increases when facing limited sources. In 2040, the total energy production is estimated at 50,000TWh per year where 35.000TWh comes from fossil fuels [1]. **Figure 1** illustrates the total energy production in the world:

The dominant production sources are fossil fuel systems (combustion, gasification, oil, and gas) that work continuously by burning the fuel in a burner or incinerator (in the case of solid fuels). The efficiency of these systems depends on Moisture contents of the fuel, temperature of flue gas and ashes, carbon conversion or percentage of the ash and flue gas, temperature gradient in the hot gas and flue gas, and heat losses inside the reactor. Moreover, thermal power plants are designed for full-load operation because reducing the flow rate decreases combustion efficiency. Thus, using thermal systems in smart grid and off-grid is constrained by efficiency loss.

Renewable energy is intermittent and presents volatility of the source. The probabilistic distribution of wind speed limits the maximum power extracted by wind farms to 30%. Solar energy uses photon energy to move the electrons (photovoltaic system) or collect the radiation heat by using mirrors (concentrated solar panels) to a tower that heats water to produce steam. To achieve a short return on investment, one think about storing the excess produced electricity.

On the other hand, hydropower is considered as renewable as its primary source (water) does not need conversion before using it. The worldwide energy production

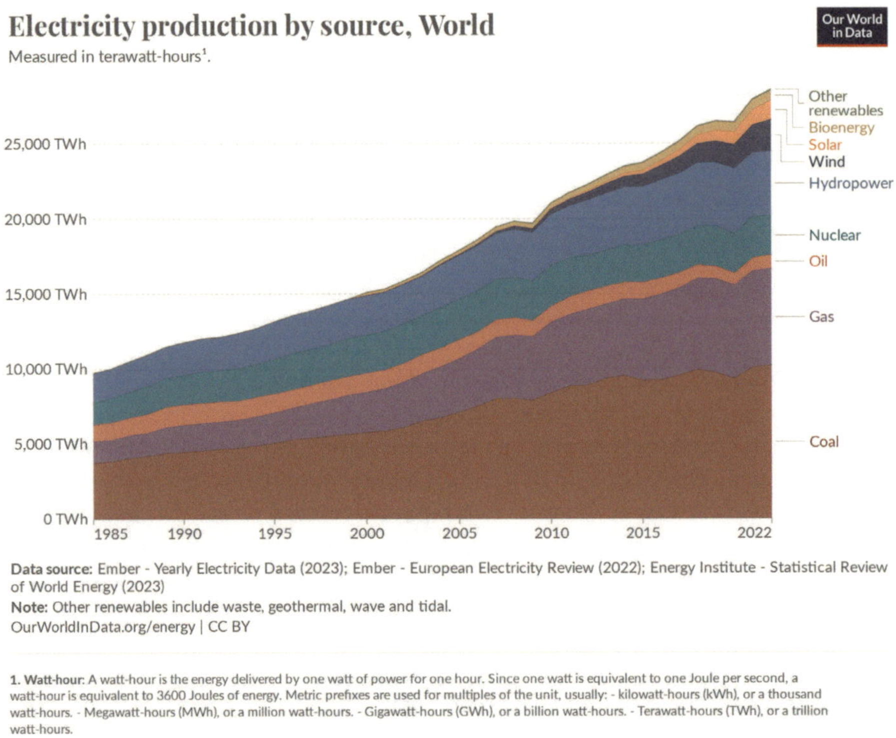

Figure 1.
The worldwide electricity production per source (TWh/year).

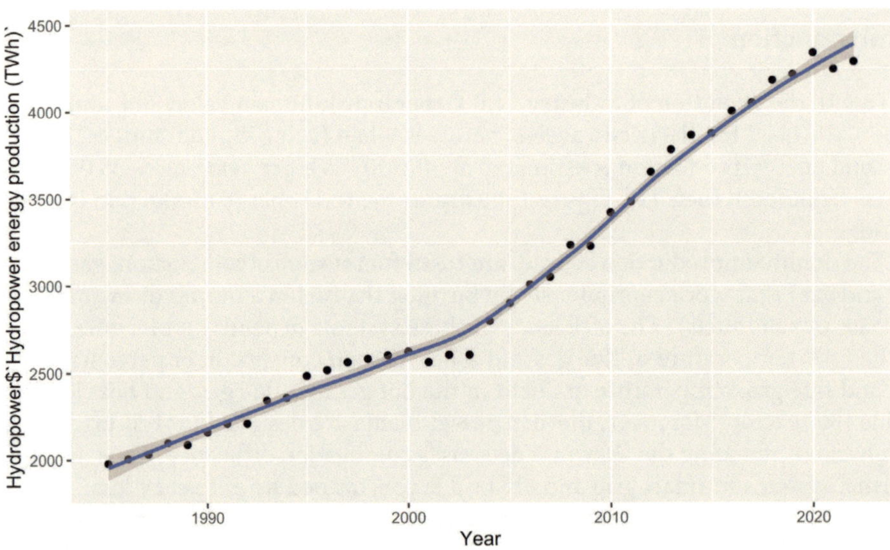

Figure 2.
Energy produced from hydropower from 1985 to 2022 (Source: Ourwoldindata.org).

from hydropower shows an increased trend from 1985 to 2022 by producing 4.200TWh. Following this trend, energy from hydropower can achieve a maximum of 6.000TWh per year in 2040 (**Figure 2**).

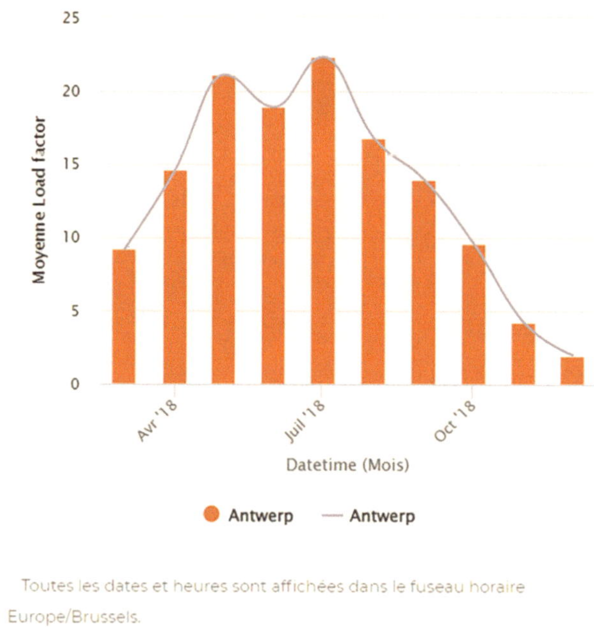

Figure 3.
Load factor of the photovoltaic production system in Antwerpen in 2018 per month.

Energy storage consists of storing the excess electricity produced when there is a lack of power supply. Different energy storage solutions exist like compressed air storage, batteries, flywheel, hydrogen, and pumped storage. From Ref. [2], by using the ranking PROMOTHEE method, hydrogen and flywheel are well placed due to the low cost and technological maturity that should be achieved for hydrogen systems.

The rule of electricity storage is to manage the difference between energy consumption and power produced. However, electricity prices change during the day. The price for peak demand is high and it is cheap in case of low demand. If we produce more than we consume during low demand, we should store the excess electricity produced to avoid grid losses.

For example, let us see the variation in photovoltaic production in Antwerpen (Belgium) in 2018 (**Figure 3**).

From May to July, we have a higher production power than other months due to the high radiance rate. If we consume less power from April to July, the excess electricity should be stored for use in the coming months (**Figure 4**).

2. Pumped storage hydroelectricity (PSH)

As it is named, PSH uses a hydraulic pump and turbine to produce electricity [**Figure 5**]. When there is an excess of the produced power, water is pumped from the lower dam to an upper reservoir using a pump. During peak load time, water is released from the upper reservoir to the lower one to produce electricity using a turbine that drives a generator.

The rotating part of the turbine or pump is called the "impeller or runner." Its design depends on turbine type and operation conditions (Q,H). Generally, PSH systems should have an altitude above 100 m to benefit from the amount of

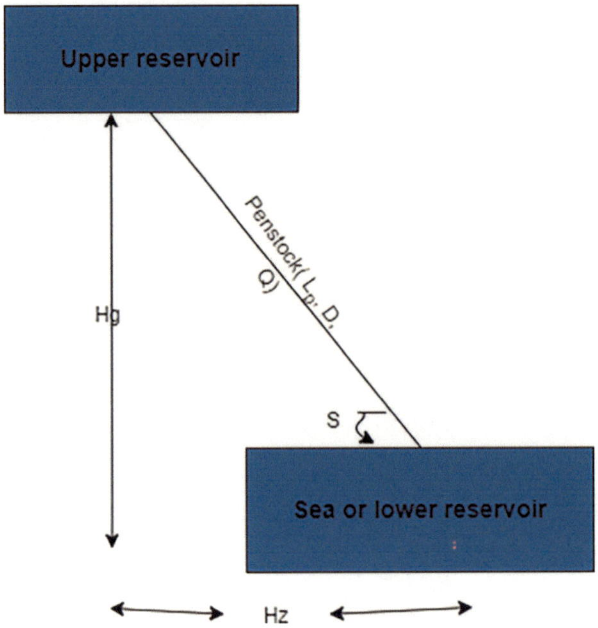

Figure 4.
Pumped storage diagram.

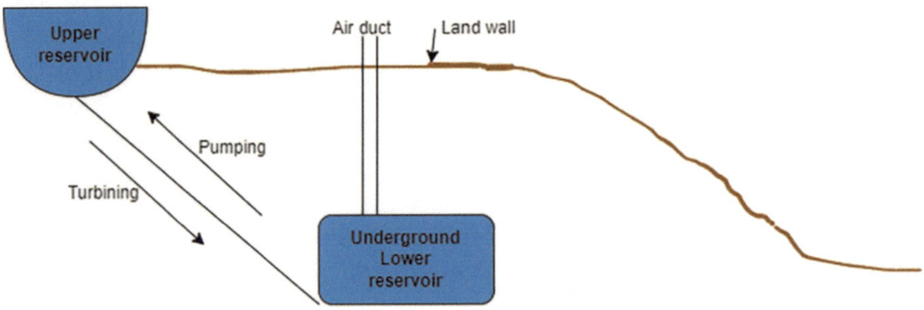

Figure 5.
Underground pumped storage [3].

electricity that should be produced. But, in other applications (small and nano) PSH, we can have a system with H less than 30 m.

The distance (Altitude) between both dams is linearly linked to the power of the system:

$$P = \eta\rho g H Q \tag{1}$$

Where H is the net latitude (m), and Q is water flow (m^3/s).

$$P = \eta\rho g(H(g) - h)Q \tag{2}$$

As H is high, stored energy becomes important with less water flow. The water is released from an upper reservoir by water intake to the penstock that carries the fluid under high pressure. After the turbine section, water is collected by a tailrace (draft tube) conduit to the lower reservoir.

The water flow is controlled by using the valve. For high-head pump storage, the use of water discharge is necessary to prevent damage from the overpressure of the fluid.

The water discharge Q depends on the energy storage duration (D_h):

$$Q = \frac{V}{D_h} \tag{3}$$

Another alternative to conventional pumped storage is underground systems. This idea was introduced as a patent by Fressden [4] in 1917. It consists of installing the upper reservoir on the earth's surface and the lower reservoir subterraneous to benefit from negative gravitation. This system can use cavities from old mines to construct the lower reservoir. One advantage of this system is to protect the conduits from atmospheric conditions (pressure, temperature, etc.). The first study was conducted by Delft Technical University [5] result to select suitable sites for a depth above 1000 m to manage wind power farms.

3. Concepts about turbine runners

The first waterwheel was developed by the French engineer Benoit Fourneyron. The American engineer James B. Francis developed the radial machines that convey water in a radial direction followed by an axial direction. This machine is widely used due to its hydraulic efficiency (flow is guided by blades).

The Pelton turbine was introduced by the engineer Lester Pelton. The flow is coming from a high-pressure pipe to a jet nozzle, which expands in atmospheric conditions. The releasing fluid impacts the blades or buckets to create the torque necessary to turn the runner (**Figures 6** and 7).

A drawing of the Francis and Pelton turbines is shown below:

Each hydraulic machine is designed for a specific condition. The machines that have the same dimensionless parameters have same efficiency.

$$\Omega_{sp} = \frac{\Omega\sqrt{Q}}{gH^{\frac{3}{4}}} \tag{4}$$

From Ref. [6], the Francis turbine has a large interval where efficiency is higher. However, changing the specific speed is due to the variation of rotation speed which

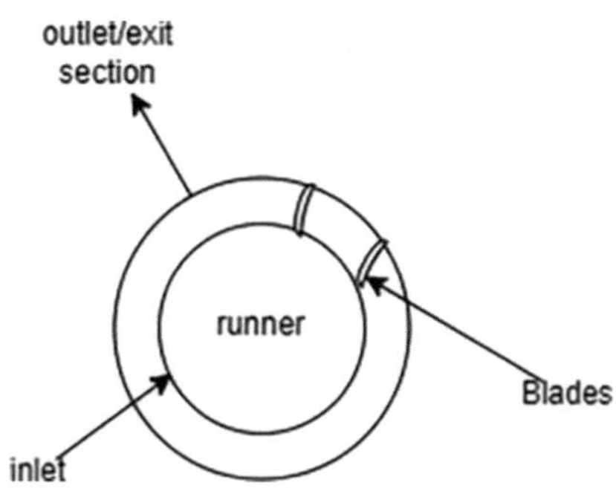

Figure 6.
Axial view of a Francis Turbine runner.

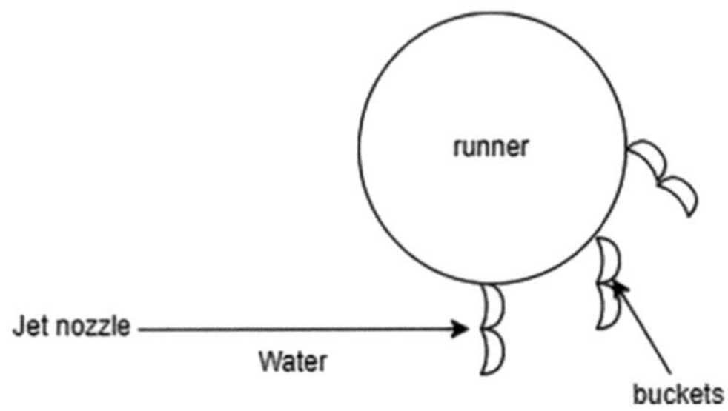

Figure 7.
Pelton turbine runner.

can be the case of part load operation. Francis's turbine can work for different delivered power with different speeds. To keep the efficiency higher, we use IGV to guide the flow in a radial direction by the curvature of the blades. The rule of IGV is to control the flow rate due to the part load condition.

The Pelton turbine is used for a low specific speed. It works at high altitudes and low specific speeds [Eq. (4)]. This type is operated for a given optimized specific speed (between 0,1 and 0,2 rad/s), and the jet nozzle is separated from the runner. If we change the runner's speed, we lose efficiency due to the torque variation by the kinetic energy of the fluid.

3.1 Francis turbine (FT)

In fluid mechanics, the absolute velocity of a rotating machine is divided into peripheral velocity and relative component.

$$\text{Absolute velocity} = \text{relative velocity} + \text{Peripheral velocity} \tag{5}$$

The FT is constituted of a volute (guiding the flow with constant velocity), Inlet Guide Vanes (IGV), and a runner. The volume conveys the liquid to the entrance of IGV. The rule of the stator part is to decrease the absolute velocity of water and increase the absolute pressure. After leaving the IGV, fluid enters the inlet section of the runner, which follows the blade curvature to decrease the pressure and reach the atmospheric value (at the outlet). The **Figure 8** shows a 3D model for the FT component set.

The blade's geometry is designed as the flow should be guided in parallel to its curvature (**Figure 9**). If we zoom (**Figure 10**), the geometry is illustrated as follows:

The flow enters the turbine set with a volute that guides the liquid to the entrance of IGV. From the inlet of the volute to its outlet, the circumferential velocity is increased with the area variation. The area variation is set concerning:

$$rC = constant \tag{6}$$

At the entrance of the blade runner, the flow has three velocity components: tangential, meridional, and relative velocities (**Figure 11**).

From (**Figure 12**), the absolute velocity C is the sum of the circumferential velocity and relative velocity:

$$\vec{C} = \vec{U} + \vec{W} \tag{7}$$

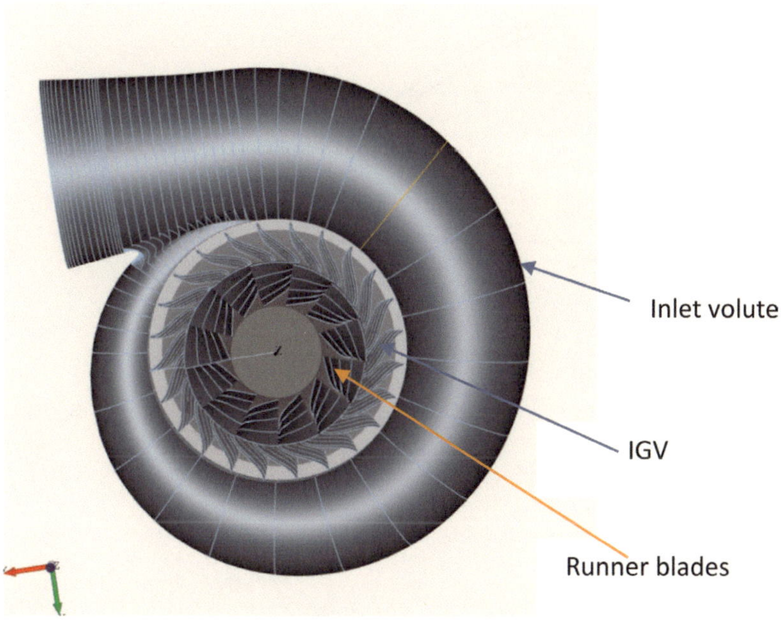

Figure 8.
3D geometry of Francis turbine sets.

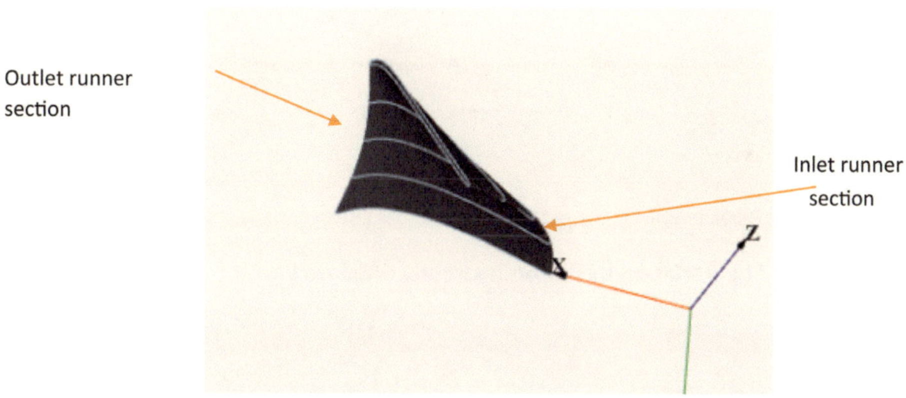

Figure 9.
3D geometry of an FT runner blade.

The blade angle β is defined as the angle between the peripheral velocity and tangential vector $\beta = \left(\vec{U}, \vec{W}\right)$. From Eq. (7), we can have:

$$C^2 = U^2 + W^2 + 2UW\cos\beta \tag{8}$$

Respectively, at the inlet and outlet, we have:

$$C_1{}^2 = U_1{}^2 + W_1{}^2 + 2U_1W_1\cos\beta_1 \tag{9}$$

$$C_2{}^2 = U_2{}^2 + W_2{}^2 + 2U_2W_2\cos\beta_2 \tag{10}$$

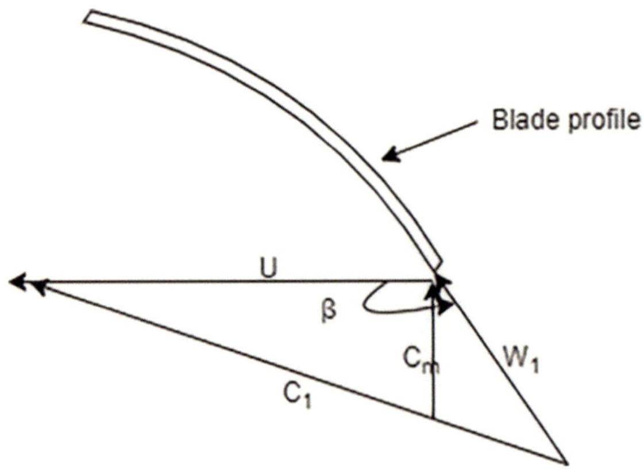

Figure 10.
Inlet triangle velocity at the runner section.

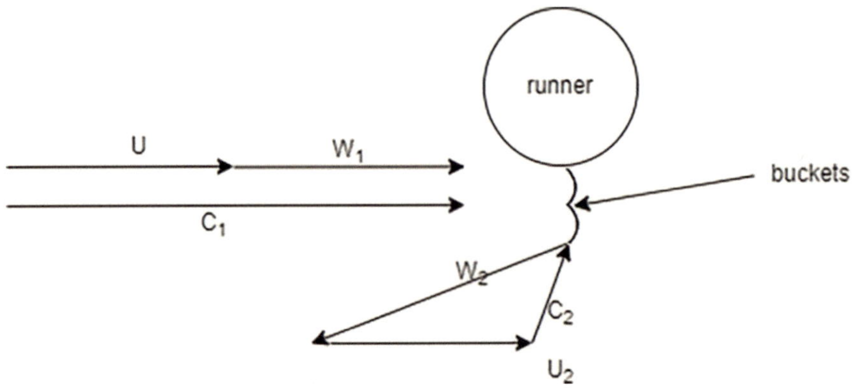

Figure 11.
Velocity triangle for Pelton turbine.

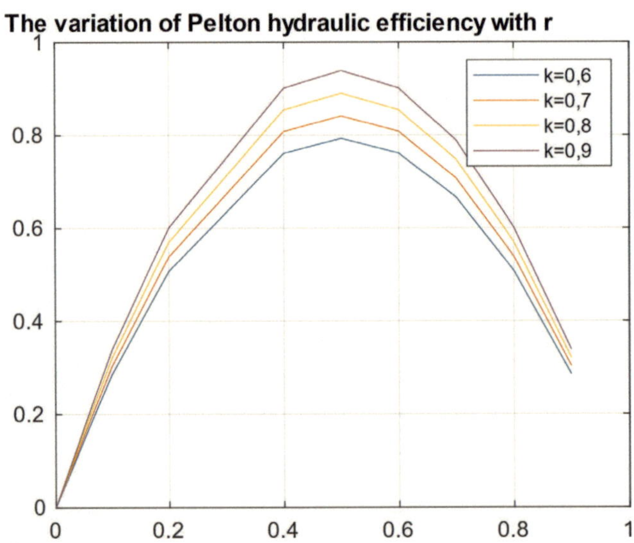

Figure 12.
Variation of Pelton hydraulic efficiency with jet speed ratio.

The torque created due to the tangential velocity is:

$$\tau = r_1 C_{u1} - r_2 C_{u2} \tag{11}$$

The power delivered by the runner is:

$$P = \tau * w \tag{12}$$

If the flow is living without swirl, Eq. (11) is simplified as:

$$\tau = r_1 C_{u1} \tag{13}$$

At the entrance of the runner, the total energy available is the sum of the pressure difference, kinetic energy due to the fluid velocity (C_1), and potential energy:

$$\rho g(H_E - \Delta H_N) = (p_2 - p_a) + \frac{1}{2}\rho {C_1}^2 + \rho g z_2 \tag{14}$$

After simplification, we have:

$$g(H_E - \Delta H_N) = \frac{(p_2 - p_a)}{\rho} + \frac{1}{2}{C_1}^2 + g z_2 \tag{15}$$

At the draft tube, the total energy available is:

$$g(H_E - \Delta H_N - \Delta H_R) - P = \frac{(p_3 - p_a)}{\rho} + \frac{1}{2}{C_3}^2 + g z_3 \tag{16}$$

The hydraulic efficiency is defined as the ratio of the total power delivered divided by the total energy available at the entrance of the runner:

$$\eta_H = \frac{P}{g H_E} \tag{17}$$

In a radial turbine (FT), the tip speed ratio (U_1/C_1) should be kept below 1 as from [Eq. (7)]. In general, the usual values are between 0.6 and 0.9 [6]. This is the advantage of FT because a large interval of this ratio means that the runner can operate with different rotational speeds and volumetric flow for high efficiency.

3.2 Pelton turbine

The Pelton turbine is different from Francis's turbines. The kinetic energy of the flow is used to rotate the Pelton wheel. It is used for very high altitudes.

At the entrance, we have:

$$C_1 = W_1 + U \tag{18}$$

At the outlet:

$$C_{u2} = U + W_2 \cos \beta_2 \tag{19}$$

The value of the blade angle (β_2) should be above 160 deg. to avoid that water will interact from one bucket to the following one.

The power delivered by the turbine is:

$$P = U_1 C_{u1} - U_2 C_{u2} = U(W_1 + U - U - W_2 \cos\beta_2) = U(W_1 - W_2 \cos\beta_2) \quad (20)$$

As $C_{u1} = C_1$ and $U_1 = U_2 = U$.

By substituting Eq. (18) into Eq. (20), we have:

$$P = (C_1 - W_1)(W_1 - W_2 \cos\beta_2) \quad (21)$$

If, we put:

$$r_v = \frac{W_1}{C_1} \quad (22)$$

The hydraulic efficiency is:

$$\eta_H = \frac{P}{\frac{C_1^2}{2}} = 2\left(r_v - r_v^2\right)\left(1 - \frac{W_2}{W_1}\cos\beta_2\right) \quad (23)$$

The variation of η_H with r_v is plotted below. It shows that the maximum efficiency is reached at $r_v = 0{,}5$.

If we suppose that:

$$r_v = \frac{W_1}{C_1} = \frac{C_1 - U}{C_1} = 1 - \frac{U}{C_1} \quad (24)$$

It means that the efficiency is maximum at jet speed ratio (U/C_1) equal to 0.5.

3.3 Penstock

Figure 13 shows the flow directions regarding the penstock section and umped storage operation mode (pumping and generation):

From the mass conservation, the mass flow rate is constant:

$$\dot{m}_1 = \dot{m}_2 \quad (25)$$

As the fluid is moving under high pressure, the speed of sound or pressure wave velocity is considered as one characteristic of the flow defined by the Eq. (26).

$$a = \sqrt{\frac{K}{\rho\left(1 + \frac{KD}{eE}\right)}} \quad (26)$$

For the elastic pipe or penstock, the ratio (D/e) depends on Poisson's coefficient [7, 8]. If we define ψ as:

$$\psi = 2(1 + v)\frac{R_o^2 + R_i^2}{R_o^2 - R_i^2} - 2\frac{vR_i^2}{R_o^2 - R_i^2} \quad (27)$$

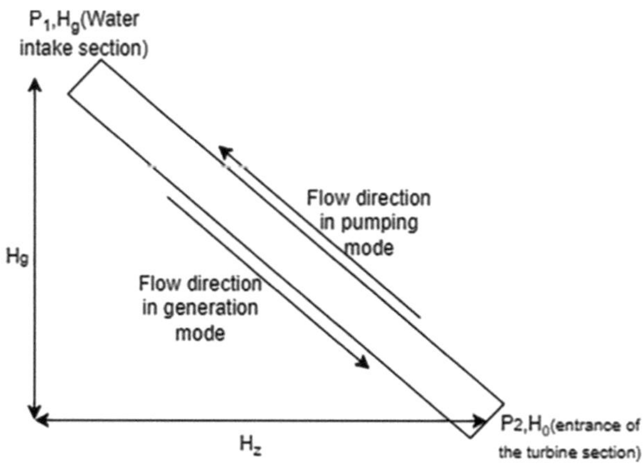

Figure 13.
Flow inside the penstock section.

Where v is Poisson's ratio for a given material. R_i and R_o are respectively the internal and external radius of the penstock. Eq. (26) can be written as:

$$a = \sqrt{\frac{K}{\rho\left(1 + \frac{\psi K}{E}\right)}} \tag{28}$$

In Okinawa pumped storage, the fluid used inside the penstock is seawater. It differs from natural water in terms of its thermophysical properties. Due to the presence of salt, the density and electrical conductivity of seawater is slightly higher than natural water. The change of density induces the deviation of the wave velocity (Eq. (27)), which can result in the variation of the volumetric flow of the liquid.

3.4 Seawater-pumped storage

Seawater-pumped storage (SWPSH) consists of using the sea as a lower reservoir instead of an artificial dam. This could help to reduce the construction and excavation costs. Seawater compared to natural water is different in terms of its thermophysical parameters (viscosity, salinity, etc.) (**Figure 14**).

The salinity of seawater causes corrosion mainly for the penstock and turbomachinery sides. The weight loss due to the corrosion is linked to the resistivity decreases of seawater.

The location of SWPSH is mostly situated near the shoreline (up to 1 km) and in the presence of cliffs. The first project was built in Japan and named Okinawa seawater-pumped storage (**Figure 15**).

Our study at the University of Brussels conducted the location of some suitable sites in Europe. As an example, the Manneville site in North France with an available surface of 28km^2 (**Figure 16**).

In North Africa, we concluded that at least 11 sites are available in Morocco with a storage capacity of up to 360MWh [2].

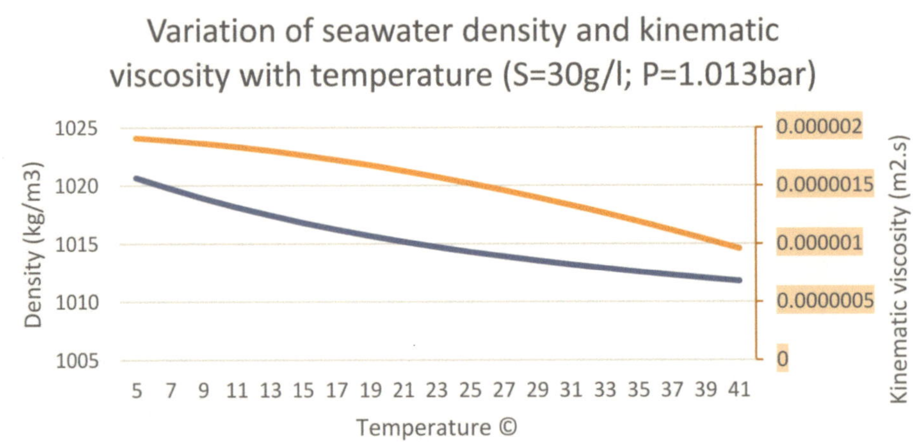

Figure 14.
Seawater properties with temperature (C).

Figure 15.
Okinawa seawater-pumped storage location in Japan.

3.4.1 Hydraulic design

Three main parameters that characterize penstock geometry are diameter (D), Length (L_p), and bends (due to the site topography). These parameters are directly linked to the hydraulic design and optimization of the water pipe. The total head losses depend on the square of fluid velocity, pipe diameter, bends coefficients, and pipe length.

$$h = fL_p \frac{V^2}{2D} + k\frac{V^2}{2g} \tag{29}$$

Figure 16.
Manneville site in North France (49°52′02″N 0°45′42 E).

The friction factor (f) is linked to Reynolds number by Colebrook white equation:

$$\frac{1}{\sqrt{f}} = -2\log_{10}\left[\frac{2.51}{R_e\sqrt{f}} + \frac{\varepsilon}{3.7D}\right] \tag{30}$$

Where ε is the relative roughness of the pipe.

$$R_e = \frac{VD}{\vartheta} \tag{31}$$

The friction factor f decreases when R_e increases. Different equations are used to calculate the penstock diameter with Warnick equation for a gross head between 60 and 100 m, and a rated power between 150 and 750 MW [9].

$$D_e = 0.71\frac{P^{0.43}}{H_g{}^{0.65}} \tag{32}$$

The erosion process consists of material losses due to solid particle attack on the surface of a metal. Commonly, the erosion rate is presented as the ratio of the eroded mass divided by the material mass (kg eroded/material mass). The impingement angle is the angle in which the particle is in contact with the internal surface of the metal (**Figure 17**).

The erosion depends on the particle size, attack angle, and particle velocity. Finnie's models [10, 11] define the erosion rate as:

$$E_r = kf(\alpha){V_p}^2 \tag{33}$$

K is a coefficient that depends on the hardness index of the material and f(α) is a coefficient that depends on the attack angle. For the ductile material, this coefficient is maximum for an angle α of 45 deg. The velocity of the particle is designed as V_p.

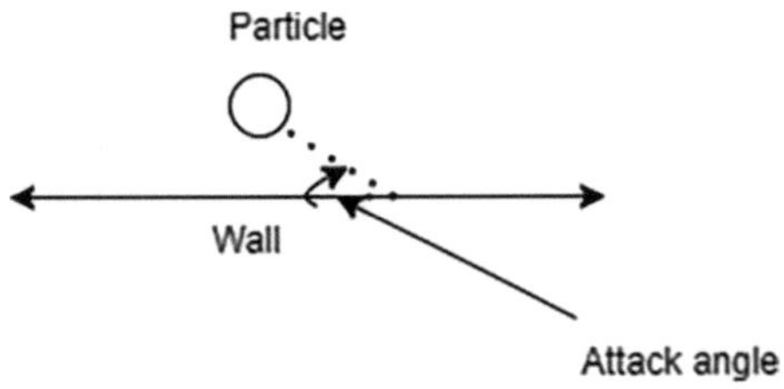

Figure 17.
Erosion mechanism.

One important parameter is the particle velocity distribution. Due to the drag forces, particle trajectory changes between medium (air, water, oil, etc.). The Lagrangian theory defines the equation to describe the particle trajectory by defining the velocity distribution in the 3D frame.

$$E_p = C_D * E_{sw} \tag{34}$$

Where C_D is the Drag coefficient, and E_p and E_{sw} are the kinetic energy of the particle and seawater (fluid), respectively.

As E_p and E_{sw} are dependent on the medium mass, then they are related to their densities. The particle path is linked to the ratio of densities. If the density of the fluid is higher than the particle, particles are more likely to follow the flow trajectory.

In the hydraulic machine, the slip factor (coefficient that measures how much the flow is guided by the blades) is linked to the blade thickness and meridional component of the absolute velocity. If the thickness loss becomes higher, flow separation may occur, which causes a drop in efficiency.

In Ref. [12], the authors concluded by studying the erosion in the blade-to-blade section of low-head pump. Erosion is high at the trailing edge where the absolute velocity is high.

The blade blockage is defined as:

$$\sigma = \frac{t}{t - \frac{2\pi r}{z}} \tag{35}$$

The variation of the meridional velocity is:

$$C_{m2} = \sigma C_{m1} \tag{36}$$

The effect of reducing blade thickness by erosion is to change blade blockage [Eq. (35)] area which causes a variation of the meridional velocity component.

3.5 Using pumps as turbine (PAT)

Nowadays, some applications need to use pumps as turbines to reduce investment costs. By this, the pump works in normal situations by pumping water from the low head to an upper reservoir. On the other hand, in reverse mode (turbine), the rotational direction is inversed. It is very difficult to predict losses in the turbine mode in volute, cut water, and casing due to the inversed flow [13].

If we assume:

$$P_T = P_P = \eta_T Q_T H_T g = \frac{Q_p H_p g}{\eta_p} \tag{37}$$

If we assume that $\eta_T\eta_p$ = 1 then, then Eq. (37) becomes:

$$\frac{Q_p H_p}{Q_T H_T} = 1 \tag{38}$$

If the case that $\eta_T = \eta_p$, then Eq. (37) becomes:

$$\frac{Q_T H_T}{Q_p H_p} = \frac{1}{\eta_p{}^2} \tag{39}$$

The head ratio is defined as:

$$h = \frac{H_T}{H_p} \tag{40}$$

$$q = \frac{Q_T}{Q_p} \tag{41}$$

At the BEP, the pump efficiency is equal to

$$\eta_p = \eta_H{}^{0.5} \tag{42}$$

From Eq. (40), as $\eta_p < 1$, the pump will work with higher heads in turbine mode. From Ref. [14], a general head ratio can be written as:

$$h = \frac{b}{\eta_p{}^a} \; where \; a, b > 1 \tag{43}$$

The triangle velocity is different in pump and turbine mode. However, when the pump operates in turbine mode, the flow has an incidence angle at the inlet section. The volute works as guide vanes.

As for PaT operations, the specific speed of the turbine is given by:

$$n_{st} = \sqrt{\eta_p} n_{sp} \tag{44}$$

Reducing the rotational speed during the turbine mode may keep the hydraulic efficiency the same as in pump mode (**Figure 18**) [15].

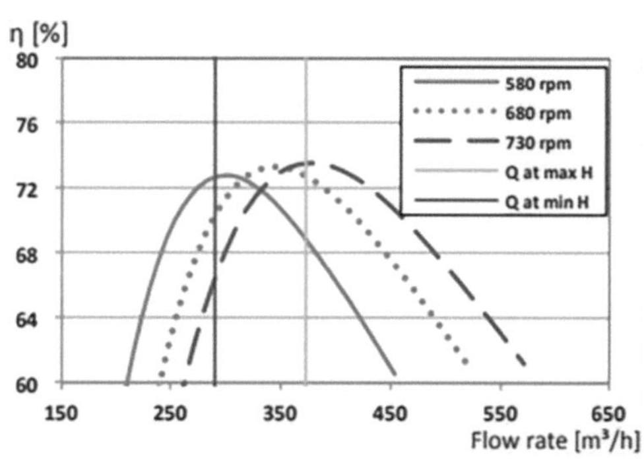

Figure 18.
The variation of PaT with RPM [15].

4. Conclusion

Pumped storage is considered a good solution to store the excess electricity compared to other storage options. The regions with high altitudes are the most appropriate sites to install PSH near the shoreline, cliffs, and mountains.

The choice of turbomachinery depends on water flow (Q), altitude (H), and operation conditions (pump generation or separate pumping). Francis's turbines are used for different loads with high efficiency. Impulse turbine is used for high altitudes with low specific speed. The design of pumped storage should have an object to reduce the installation cost and minimize the hydraulic head losses (kept below 1%).

The material losses due to the erosion process are related to the particle velocity, hardness index, and attack angle. In the blade-to-blade section, thickness loss is linked to the flow absolute velocity.

Pumped storage is a suitable solution to store electricity, especially using seawater. The development of pump as turbine (PAT) technology makes nano application of PSH possible without changing the pump in an existing installation.

Acknowledgements

This book chapter is dedicated to some persons who allowed me to learn new things and grow. Firstly, I address this work to my former Boss "Thierry Jehin" who teaches me patience for work and success. Secondly, to my brother (El Habib El Mansour) and my parents who encouraged me to participate in this book project and continue my scientific research in energy storage.

To my Professor Patrick Hendrick at ULB who teaches me every day and pushes me to do my best to get good results and achieve my scientific goals.

Conflict of interest

The authors declare there is no conflict of interest.

Abbreviations

h	head losses (m)
H	net head (Altitude) between both reservoirs (m)
H_g	gross head between both reservoirs = H-h
Q	fluid flow (m^3/s)
V	reservoir volume (m^3)
D_h	energy duration (s)
SW	seawater
C_m	Meridional velocity (m/s)
C_u	tangential velocity (m/s)
C	absolute velocity (m/s)
U	peripheral velocity (m/s)

Author details

Jamal El Mansour
ATM, Universite Libre de Bruxelles, Brussels, Belgium

*Address all correspondence to: elmansour.j@ucd.ac.ma

References

[1] Lempérière F. (Hydrocoop). Hydropower plants may increase from 1 300 GW to 5 000 GW. Available from: http://wwwhydrocooporg/hydropower-plants-may-increase-from-1-300-gw-to-5-000-gw/2018.

[2] Elmansour J, Hajjaji A, Belhora F, Hendrick P. Location of seawater pumped storage hydropower plants: Case of Morocco. Materials Today Proceedings. 2022;**66**:45-57. DOI: 10.1016/j.matpr.2022.03.145

[3] Bodeux S, Pujades E, Orban P, Brouyère S, Dassargues A. Interactions between groundwater and the cavity of an old slate mine used as lower reservoir of an UPSH (underground pumped storage hydroelectricity): A modelling approach. Engineering Geology. 2017;**217**:71-80. DOI: 10.1016/j.enggeo.2016.12.007

[4] Reginald A. Fessenden. System of storing power. 1917

[5] Min APN. Ondergrond Pump Accumulatie Centrale. Department Hydraulic Engineering, TU Delft; 1984. Available from: https://resolver.tudelft.nl/uuid:a87a1047-fbed-4f39-9c86-7e829980a216

[6] Dixon SL, Hall CA. Chapter 9 - Hydraulic turbines. In: Dixon SL, Hall CA, editors. Fluid Mechanics and Thermodynamics of Turbomachinery. 7th ed. Butterworth-Heinemann; 2014. pp. 361-418. ISBN 9780124159549

[7] Chaudry MH. Applied Hydraulic Transient. New York: Van Nostrand Reinhold; 1979

[8] Nicolet C. Hydroacoustic modelling and numerical simulation of unsteady operation of hydroelectric systems. Ecole Polytechnique de Lauzanne. [Doctoral thesis]. 2007. pp. 80-100

[9] Kamran M. Chapter 6 - Hydro energy. In: Kamran M, Fazal MR, editors. Renewable Energy Conversion Systems. Academic Press; 2021. pp. 193-219. DOI: 10.1016/B978-0-12-823538-6.00007-5. ISBN 9780128235386

[10] Thapa B. Sand Erosion in Hydraulic Machinery. Fakultet for ingeniørvitenskap og teknologi; 2004. Available from: http://hdl.handle.net/11250/231204

[11] Sangal S, Singhal MK, Saini RP. Hydro-abrasive erosion in hydro turbines: A review. International Journal of Green Energy. 2018;**15**: 232-253. DOI: 10.1080/15435075.2018.1431546

[12] El Mansour J, Hendrick P, Hajjaji A, Belhora F. The Effect of the Erosion on the Hydraulic Machine's Performances. Spring Nature; n.d. DOI: 10.1007/s41207-024-00480-9

[13] Morabito A, Steimes J, Hendrick P. Pumped hydroelectric energy storage: A comparison of turbomachinery configurations. In: Sustainable Hydraulics in the Era of Global Change. Proceedings of the 4th IAHR Europe Congress, CRC Press; 2016. pp. 261-268. DOI: 10.1201/b21902-48

[14] Carravetta A, Derakhshan Houreh S, Ramos HM. Pumps as Turbines. Cham: Springer International Publishing; 2018. DOI: 10.1007/978-3-319-67507-7

[15] Morabito A, Steimes J, Bontems O, Al Zohbi G, Hendrick P. Set-up of a pump as turbine use in micro-pumped hydro energy storage: A case of study in Froyennes Belgium. Journal of Physics Conference Series. 2017;**813**:012033. DOI: 10.1088/1742-6596/813/1/012033